Basic
Mathematics
for
Biochemists

Basic Mathematics *for* Biochemists

Second edition

Athel Cornish-Bowden

Directeur de Recherche,
Centre National de la Recherche
Scientifique, Marseilles

OXFORD
UNIVERSITY PRESS

OXFORD
UNIVERSITY PRESS

Great Clarendon Street, Oxford OX2 6DP

Oxford University Press is a department of the University of Oxford.
It furthers the University's objective of excellence in research, scholarship,
and education by publishing worldwide in

Oxford New York

Athens Auckland Bangkok Bogotá Buenos Aires Calcutta
Cape Town Chennai Dar es Salaam Delhi Florence Hong Kong Istanbul
Karachi Kuala Lumpur Madrid Melbourne Mexico City Mumbai
Nairobi Paris São Paulo Singapore Taipei Tokyo Toronto Warsaw

with associated companies in Berlin Ibadan

Oxford is a registered trade mark of Oxford University Press
in the UK and in certain other countries

Published in the United States
by Oxford University Press Inc., New York

First edition published 1981 © Chapman and Hall
Second edition published 1999 © Athel Cornish-Bowden 1999

A catalogue record for this book is available from the British Library

Library of Congress Cataloging in Publication Data
ISBN 0 19 850217 6 (Hbk)
ISBN 0 19 850216 8 (Pbk)

Typeset by Newgen Imaging Systems (P) Ltd., Chennai, India

Printed in Great Britain
on acid-free paper by
Biddles Ltd., Guildford & King's Lynn

To Isadora

Preface

Some teachers of biochemistry think it is positively beneficial for students to struggle with difficult mathematics. Such teachers are less numerous than they were when the first edition of this book appeared nearly 20 years ago, but they still exist. They are wrong, in any case, because students choose biochemistry courses in order to study biochemistry, and they ought not to be burdened with more mathematics than is actually needed for a proper understanding of biochemistry. This includes physical chemistry, of course, because physical chemistry is an essential component of biochemistry, and a biochemist who knows no physical chemistry is no biochemist. These considerations have largely determined the topics that are covered in this book. Some, like the use of logarithms, are given heavy emphasis, because they have a central role in the subject or because they often cause problems for students. Others, such as trigonometry, are largely ignored, because regardless of how important they may be in mathematics as a whole they find little application in biochemistry, especially elementary biochemistry. The words 'for biochemists' in the title of the book are thus essential: it is not a guide to mathematics for science students in general, or even for students in subjects like chemistry or genetics that also require mathematics but whose needs do not coincide with those of biochemists.

Having decided what to include, there were two other questions that had to be decided: how elementary should the treatment be, and in what order should the topics be dealt with. The first of these was quite easy to decide, though it will be for the reader to judge how successful I have been: even though almost all students using the book will have had at least some previous mathematical experience, I have tried to assume an absolute minimum of prior knowledge (other than simple arithmetic), and to explain everything. On the one hand, it is much easier to skip rapidly through information that is not needed than it is to fill in information that is missing or dealt with too briefly. On the other hand, even if the material has been covered already, it may not be remembered very thoroughly: biochemistry students may not have realized when they first studied equation solving or graph drawing how much of it they were going to need later. To summarize this paragraph, therefore, I have not followed the advice of one reviewer of the first edition, who felt that it was written at too basic a level for most of his students: I suspect, in any case, that he was making a statement more about his own teaching than about what his students could really follow. On the contrary, I have moved in the opposite direction: this book is longer than its predecessor, because it includes more basic material, it explains more—leaving fewer steps to be 'understood'—and it develops the subject more slowly throughout.

The question of the order in which to present the different topics was more difficult to decide, because a sequence that never referred to anything that would be dealt with later would require a much longer book, with many more chapters, and a time scale for years rather than months of teaching it. It would also be unrealistic, because even a student who has largely forgotten what a logarithm is will not have forgotten so completely that it cannot be mentioned at all before it is defined. So, although I have tried mainly to work forwards, so that the later chapters build on the earlier ones, some of the more advanced topics nonetheless come earlier than some of the simpler ones, most notably in the case of equation solving: solving simple linear equations could certainly come quite early in the book, if it were not for the need to keep it logically with some aspects of equation solving that require a prior knowledge of calculus.

I should like to thank Howard Stanbury and Liz Owen at Oxford University Press for their encouragement to proceed, and Lisa White for many useful comments both on the original outline and on the first draft. Detailed reading by Marilú Cárdenas of the sections added for the first time in this edition, were very helpful, and her vigilance in checking the solutions to the problems allowed numerous errors to be avoided.

December 1998   Athel Cornish-Bowden
                                                           Marseilles

Note: Any typographical errors detected after this book is printed will be listed on the world wide web at the following URL. http://ir2lcb.cnrs-mrs.fr/~athel/basmaths.htm

Contents

CHAPTER 1

Basic ideas

1.1 **Introduction**

In this chapter we introduce some basic ideas that nearly all readers will already be familiar with. Some will be examined again in more detail later in the book, but even those that are not mentioned again specifically underlie much of the more advanced material, and so readers seeking to improve their mathematical knowledge and understanding need to feel confident that all of the basic ideas have been mastered.

Let us consider the following questions listed in **Box 1.1** as an introductory guide, which are all intended to be solved without the use of a calculator. The last four (18–21) can all be solved either by writing out all the possibilities and counting them or by using appropriate formulas. Ideally, you should be confident of being able to do them *both* ways: writing out all the possibilities is hardly practical for more complicated questions of the same type that may arise in the study of protein or nucleic acid sequences, whereas applying a formula blindly is dangerous if one has so little idea of what it means that one cannot solve even the simplest problem any other way.

If all of these questions seem trivially easy to answer then you can probably proceed to the next chapter immediately (though it might be wise to check the answers first: they are listed at the end of the chapter). Questions 1–5 are all concerned with elementary manipulations of fractions and the relationship between fractions and decimal numbers. If any of these give problems it is important to follow through the relevant part of this chapter, because the relationships will be needed elsewhere in the book and they will not be discussed explicitly again. The algebraic questions are here mainly concerned with understanding the algebraic symbolism: equation solving, etc., will be dealt with later in the book, but not at

Box 1.1 **Revision questions**

1 How can $\frac{14}{35}$ be written as an equivalent fraction involving smaller numbers in the numerator and denominator?

2 What is the value of $\frac{2}{3} + \frac{3}{5}$ (expressed as a fraction)?

3 What is the value of $\frac{2}{3} \times \frac{3}{5}$, expressed as the simplest possible fraction?

4 What is the value of $\frac{2}{3} \times \frac{3}{5}$, expressed as a decimal number?

5 What would 0.75 be if written as a simple fraction?

6 What would 8.37×10^{-3} be if written as an ordinary decimal number?

7 If $x = 3$, $y = 2$, and $z = 5$, what is the value of $(x+y)(y-z)$?

8 If $2x + 7 = 13$, what is the value of x?

9 Write 3^3 as a multiplication expression: what is its value?

10 If $7x + 3 = 3y + 1$ and $x = 4$, what is the value of y?

11 The surface area of a sphere is double the product of the radius and the circumference at the equator. Define symbols for the three quantities mentioned and write down the relationship between them as an algebraic expression. If the circumference at the equator is $2\pi r$ (where r is the radius), what is the surface area in terms of π and r?

12 Which of the numbers 2, 7, 13, 15 and 23 is not a prime?

13 What are the prime factors of 42?

14 What is the value (as an algebraic expression) of $(a+b)(a-b)$?

15 What are the factors of $(x^2 - y^2)$?

16 Give the value of 4! (where the ! is a mathematical symbol, not a mark of surprise).

17 If a dissociation constant has a value of 3.8 mM, what is its reciprocal (the corresponding association constant), expressed in M^{-1}?

18 How many different ways are there of arranging the four letters A, C, G and T in sequence?

19 How many different ways are there writing four-letter sequences using only the letters A, C, G and T?

20 How many different ways are there of choosing two different letters out of A, C, G and T (taking no more than one of each, and without regard to order)?

21 How many different two-letter sequences can be written with A, C, G and T (taking no more than one of each)?

Notes and solutions will be found in Section 1.11 (p. 26)

such an elementary level as in this chapter. Question 11 is an example of a problem that occurs very frequently in all kinds of science, but is often inadequately covered in elementary mathematics courses: it is one thing to be able to do elementary algebraic manipulations of an expression if it is presented at the outset as an algebraic expression, but it can be more difficult when the basic information is not initially presented as algebra, so that one has first to translate a statement in everyday descriptive language into terms where algebra can be used.

Question 17 is an example of a problem of converting between units of the kind that occur frequently in all experimental sciences. A quantity is expressed in one way, but needs to be expressed differently for use in later manipulations or comparison with other measurements. This type of conversion is in principle almost trivial, but it is striking how often it is done incorrectly, even by working professional scientists—only the day before writing this paragraph I was reading a paper submitted to a scientific journal in which a calculated association constant was wrong by a factor of around 1 000 000 000 000 000.

For the last four questions it is important to grasp the distinction between a permutation (where the order matters) and a combination (where it does not). If you feel that 18 and 19, or 20 and 21, are the same question, then you need to review this point.

1.2 **Manipulating algebraic expressions**

A high proportion of the work in applying mathematics to scientific problems consists of manipulating algebraic expressions. For example, expressions like $(a+b+c^2)$ can be subjected to most of the same operations that can be applied to numbers, in most cases just as easily.

Addition of two simple sums of terms just produces a single sum containing all the terms:

$$(a+b+c^2)+(5x-y)=a+b+c^2+5x-y$$

and if any terms of the same kind occur in both sums they can be combined:

$$(a+b+c^2)+(5x-b)=a+b+c^2+5x-b=a+c^2+5x$$

Subtraction is handled in exactly the same way:

$$(a+b+c^2)-(5x-y)=a+b+c^2-5x+y$$

$$(a+b+c^2)-(5x-b)=a+b+c^2-5x+b=a+2b+c^2+5x$$

The only important things to remember are that a minus sign applied to a bracket applies to each individual term inside the bracket and that a minus sign actually means 'change the sign', so putting a minus sign in front of a negative quantity produces a positive quantity. (More detail about how minus signs are used and how to handle negative numbers is given in **Box 1.2**, which also includes some ideas to be developed further in Chapter 2.)

Multiplication is also conceptually straightforward. It is sufficient to remember that when two sums are multiplied together the result is a sum in which *every* term in the first sum is multiplied separately by *every* term in the other:

$$(a+b+c^2)\,(5x-y)=a(5x-y)+b(5x-y)+c^2(5x-y)$$

$$=5ax-ay+5bx-by+5c^2x-c^2y$$

Box 1.2 **Handling negative numbers**

A minus sign changes the sign of whatever follows it. There are actually two kinds of minus signs, written in exactly the same way except that the *unary minus sign* is written snugly against the number or expression it qualifies whereas the *binary minus* sign has a small amount of space on each side of it. The unary minus sign acts on a whole expression and changes its sign: from positive to negative if it is positive to start off with, or from negative to positive if it is negative to start off with:

-4

$-(-4)=+4$

$-(-(-4))=-(+4)=-4$

etc.

The binary minus sign is written between two expressions and acts as an instruction to subtract the second from the first, or to change the sign of the second and then add them (which is a different way of saying the same thing):

$4-3=1$

Adding a negative number is just subtraction: $4+(-3)=4-3=1$

Subtracting a negative number is equivalent
  to adding the corresponding positive number: $4-(-3)=4+3=7$

Multiplying by a negative number is equivalent
  to multiplying and changing the sign of the
  result: $4\times(-3)=-(4\times 3)=-12$

Dividing by a negative number is similar: $\dfrac{4}{-3}=\dfrac{-4}{3}=-1.333$

Taking the *logarithm* of a negative number is
  not allowed $\log(-3)$ is **not defined**

Taking the *antilogarithm* of a negative number
  is equivalent to taking the reciprocal of the anti-
  logarithm of the corresponding positive number: $\text{antilog}(-3)=\dfrac{1}{\text{antilog}(3)}$

It is often best not to carry out such multiplications until one has to, because the 'structure' of an algebraic expression is often easier to recognize if it is written as a product of factors $(a+b+c^2)\,(5x-y)$ than if it is multiplied out.

......

Example 1.1 Removing brackets
Evaluate $(x+3y+4)(2+5x)$.

Every term in the first bracket is multiplied by every term in the second:

$x(2+5x)=2x+5x^2$
$3y(2+5x)=6y+15xy$
$4(2+5x)=8+20x$

and the three results are added together:

$2x+5x^2+6y+15xy+8+20x$

This is correct, but it contains two terms of the same sort ($2x$ and $20x$), and so these can be combined:

$22x+5x^2+6y+15xy+8$

and the terms rearranged systematically:

$5x^2+22x+15xy+6y+8$

......

Division is the exception among the four standard arithmetic operations: it is not usually possible to simplify a ratio by a simple mechanical process, so an expression such as $(a+b+c^2)/(5x-y)$ usually has to be left just like that. Simplification is possible, however, if one can recognize a common factor in the numerator (top part of a fraction) and denominator (bottom part). In the most trivial case of a pure number as factor, this may be obvious:

$$\frac{4x-6y}{10a+2b}=\frac{\cancel{2}(2x-3y)}{\cancel{2}(5a+b)}=\frac{2x-3y}{5a+b}$$

but usually one is not so fortunate. There are a few types of expressions where algebraic factors are sufficiently easy to recognize that it is useful to be familiar with them. First, consider the square of any algebraic sum:

$$(x+y)^2=x^2+xy+xy+y^2=x^2+2xy+y^2$$

or of any algebraic difference:

$$(x-y)^2=x^2-xy-xy+y^2=x^2-2xy+y^2$$

The only difference between these two results is the middle term, which is positive in the first case and negative in the second. Note in particular that the sign of y^2 is positive in *both* cases: if you do not immediately understand why this is so you need to read the early part of this section again. These two types of expressions occur often enough that one ought to be able to recognize their forms at sight: a three-term expression in which the first and last terms are squares and the middle term is plus or minus twice the product of the quantities squared. This knowledge then allows one to recognize immediately that an expression such as

$$\frac{4a^2 - 6ab + 9b^2}{4a - 6b}$$

can be simplified, because the numerator is a perfect square:

$$\frac{4a^2 - 6ab + 9b^2}{4a - 6b} = \frac{(2a - 3b)^2}{2(2a - 3b)} = \frac{(2a - 3b)(2a - 3b)}{2(2a - 3b)} = \frac{2a - 3b}{2} = a - 1.5b$$

One other case that is worth being able to recognize is the result of multiplying any sum of two expressions by the difference between the same two expressions:

$$(x + y)(x - y) = x^2 - xy + xy - y^2 = x^2 - y^2$$

This result is called the *difference between two squares*, and is always factorizable. Thus, for example,

$$\frac{x^2 - y^2}{x^2 + 2xy + y^2} = \frac{(x + y)(x - y)}{(x + y)^2} = \frac{(x + y)(x - y)}{(x + y)(x + y)} = \frac{x - y}{x + y}$$

Example 1.2 A simple factorization

Factorize $[4p^2 - 25(q + r)^2]$.

$4p^2 = (2p)^2$ is a perfect square, and $25(q + r)^2 = [5(q + r)]^2$ is also a perfect square, so the expression as a whole is the difference between two squares and has, as factors, the sum of the two terms

$$2p + 5(q + r) = 2p + 5q + 5r$$

and their difference:

$$2p - 5(q + r) = 2p - 5q - 5r$$

The original expression can thus be factorized as

$$(2p + 5q + 5r)(2p - 5q - 5r)$$

It is perhaps worth noting that xy can equally well be written yx, and proceeding in a completely mechanical way it might seem more natural to write the second term in the expansion of $(x + y)(x - y)$ as $-xy$ and the third as $+yx$. In this case it should still be obvious that they cancel, but in a more complicated expression involving more terms it might not be so obvious. It is often wise, therefore, to write the variables in a product not in the order they appear but in some systematic order, for example numerical values first followed by algebraic variables in alphabetical order. Whether this is a good idea or not will depend on the context: 'yes' if the method is clear and the aim is to be able to recognize terms of the same kind, 'no' if it is more important for the reader to understand where a result has come from. In extreme cases you may well want to do both: first list the variables in the order they appear, then rearrange them into an order that facilitates proceeding to the next step.

1.3 **Fractions**

Manipulating fractions is a very familiar activity for everyone in the context of money, especially small amounts of money; at least in the case of American money even the names of the common coins are derived from their origins as fractions: a cent is a hundredth of a dollar, a dime is a tenth of a dollar (*dime* being an old word meaning a tenth, though it has disappeared from the language in other uses), a quarter is a quarter of a dollar, a half-dollar is—not surprisingly— half of a dollar; the only exception is the nickel, whose name reflects what it is made of and not its value as a twentieth of a dollar. Even a small boy with no particular mathematical skills knows whether he has been cheated if he receives seven cents, a nickel, three dimes and two quarters in change after paying two dollars for a magazine worth 85 cents. Yet the same question, expressed in terms of fractions:

$$\text{Is } 2 - \frac{85}{100} \text{ more than, less than or the same as } \frac{7}{100} + \frac{1}{20} + \frac{3}{100} + \frac{2}{4}?$$

can seem quite formidable. The reason we do not find this formidable with money is that we instinctively ignore the fraction-suggesting names but just do the whole calculation in terms of cents:

$$\text{Is } \frac{200}{100} - \frac{85}{100} \text{ more than, less than, or the same as } \frac{7}{100} + \frac{5}{100} + \frac{30}{100} + \frac{50}{100}?$$

which simplifies immediately to a question of comparing the value of $200 - 85 = 115$ cents with $7 + 5 + 30 + 50 = 92$ cents.

There is actually a second reason why we do not find this sort of calculation a problem when working with small amounts of money. However, it has nothing to do with fractions as such, but has much more general implications for scientific calculations, so we shall defer discussion to Section 1.10.

In calculations with coins there is always a natural base unit of one cent, because all coins are integer multiples of 1¢. In other calculations with fractions we apply the same principle, but the base unit is not usually $\frac{1}{100}$; instead, we have to find it. To find the smallest possible denominator we need to know the prime factors of the denominators of the separate fractions in the sum, but as long as we do not mind if we sometimes have to work with larger numbers than strictly necessary we can just multiply all the different denominators together.

To calculate

$$\frac{2}{15} - \frac{1}{6}$$

for example, it is useful to realize that 15 and 6 have a common factor of 3, but it is no great disaster to ignore this, and just multiply 15×6 to get 90. Then

$$\frac{2}{15} - \frac{1}{6} = \frac{2 \times 6}{15 \times 6} - \frac{1 \times 15}{6 \times 15} = \frac{12}{90} - \frac{15}{90} = -\frac{3}{90} = -\frac{1}{30}$$

As we ignored the factor of 3 at the beginning we had to divide numerator and denominator by 3 at the end, but the result was just the same as we would have got by realizing at the outset that 15 and 6 are both factors of 30:

$$\frac{2}{15} - \frac{1}{6} = \frac{2 \times 2}{15 \times 2} - \frac{1 \times 5}{6 \times 5} = \frac{4}{30} - \frac{5}{30} = -\frac{1}{30}$$

This sort of approach works just as well for algebraic expressions as it does for numbers. Consider for example the expression

$$\frac{x-3}{x+5} + \frac{x-2}{x+3}$$

To evaluate it, we first recognize that we can only add the numerators if we make the denominators equal, which we can do by multiplying both numerator and denominator of each fraction by the denominator of the *other* fraction:

$$\frac{x-3}{x+5} + \frac{x-2}{x+3} = \frac{(x-3)(x+3)}{(x+5)(x+3)} + \frac{(x-2)(x+5)}{(x+3)(x+5)} = \frac{(x-3)(x+3)+(x-2)(x+5)}{(x+5)(x+3)}$$

For the purposes of this chapter we can stop there, but if we were really interested in knowing the value of this expression we should continue by carrying out the three multiplications and adding the corresponding terms in the numerator. In practice we should also usually refrain from writing the middle part of this equation, taking it as being 'understood' that we can proceed straight from the expression on the left to the one on the right. This process is called *cross-multiplication*, because it involves multiplying each numerator by the denominator that is found across the sign (plus or minus) that is between the two fractions. Such cross-multiplication is always valid, and it is the middle part of the last equation that explains why it is valid.

There is an important similarity between equations and fractions on the one hand, and an important difference on the other. They merit closer examination, to avoid confusion, and are discussed in Section 1.6. First, however, we need to examine the relationship between fractions and decimal expressions.

1.4 **Converting between fractions and decimals**

Although fractions often arise during the process of a calculation, it is quite rare in science to express a number as a fraction when writing down the final result. Instead it is much more common to write a non-integer as a *decimal*, so we need to be clear about the relationship between a decimal and a fraction. In fact, a decimal is just a short-hand way of writing a number as the sum of a series of fractions with increasing powers of ten in the denominators. For example, the number 13.354 is just a short-hand way of writing

$$13 + \frac{3}{10} + \frac{5}{100} + \frac{4}{1000}$$

It follows that it is trivially easy to convert a decimal into a fraction. First it is expanded as above. Then all the terms are multiplied in numerator and denominator by whatever power of 10 is necessary to give them the same denominator as the last term:

$$\frac{13000}{1000} + \frac{300}{1000} + \frac{50}{1000} + \frac{4}{1000}$$

so the expression can be written as a single fraction

$$\frac{13000 + 300 + 50 + 4}{1000}$$

and the numerator terms can be added

$$\frac{13354}{1000}$$

and any common factor between numerator and denominator (2 in this example) is cancelled:

$$\frac{6677}{500}$$

This procedure follows the logic of the conversion, but is more cumbersome than is really necessary. Notice that before the final cancellation the numerator was just the original number with the decimal point removed, and the denominator contained as many zeroes after the 1 as there were digits after the decimal point in the original number. This shorter route can always be followed. For example,

$$-7.353 = \frac{-7353}{1000}$$

and so on.

Proceeding in the opposite direction, i.e. converting a fraction into a decimal, is both less trivial and more often necessary: it is quite rare in science that we really want to express a number like -7.353 as a fraction, but it often happens that we need to express a number like $\frac{17}{32}$ as a decimal. Fortunately, with the universal availability of cheap electronic calculators this has become equally trivial: it is sufficient to use the calculator to divide the numerator by the denominator and the result appears:

$$\frac{17}{32} = 0.53125$$

The only question that then remains to be decided is whether to leave the result like that or to omit some of the digits. This becomes especially necessary if there are infinitely many digits:

$$\frac{2}{13} = 0.1538461538\ldots$$

In a scientific application the numerator and denominator will usually be the results of some measurements (rather than, for example, derived from a theoretical analysis) and in this case they are certainly not known exactly, and probably are not known to better than 1 in 1000. It is usually absurd, therefore, to express the value of the fraction very precisely, and it is better to *round* the result so that only three significant figures remain (we shall examine what is meant by the term "significant figures" in Section 2.8): we would write 0.53125 as 0.531 and 0.1538461538... as 0.154. Notice that we do not just retain the first three non-zero digits from the original number (which would be *truncation* rather than rounding), but we examine the first of the digits to be dropped: if this is in the range 0 to 4 (as 2 in 0.53125) then the preceding digit is left as it is; but if it is in the range 5 to 9 (as 8 in 0.1538461538...) then the preceding digit is increased by 1.

It is normally best to leave rounding until the end of a calculation, as rounding too early may result in the loss of essential information. This is considered in more detail in Section 2.8, together with discussion of how much precision ought to be retained in a final result.

Calculation of a *percentage* is in essence just a special case of converting a fraction to a fraction with a denominator of 100. For example, 60 percent, usually written with the special symbol % as 60%, just means $\frac{60}{100}$ or 0.6, and 60% of 5 is just 0.6×5, or 3. The terminology is very often applied to *changes*, so we might speak of an increase from 17.5 to 21.0 as a 20% increase, because the simple difference, 3.5, is 20% of 17.5. All of this is simple and straightforward, but there is a subtlety that has to be taken into account when calculating percentage changes: they must always be calculated with respect to the state before the change. Thus, even though increasing 17.5 by 20% gives 21.0, decreasing 21.0 by 20% gives 16.8, because the change is 20% of 21.0, or 4.2, not 20% of 17.5. Similarly, increasing a quantity twice by 10% is not the same as increasing it once by 20%: taking 17.5 as an example, as before, the first 10% increases it by 1.75, to 19.25, and the second increases it by 10% of 19.25, i.e. by 1.925, to 21.175. A similar point applies to comparisons expressed in percentages: if we say that 11 is 10% larger than 10 or that 19 is 5% less than 20 the percentage is always calculated with respect to the value that follows the 'than'. Thus 3, is 50% larger than 2, but 2 is 33% less than 3.

1.5 **Relating numbers to powers of 10**

A numerical value in science is often written as a number in the range 1–10 multiplied by an appropriate power of 10, e.g. the number 3110 would most commonly be written as 3.11×10^3. We shall defer discussion of the reasons for doing this until Section 3.1, and here we shall just examine the mechanics of the conversion: what steps are necessary to go from 3110 to 3.11×10^3 or *vice versa*? For this purpose we can regard the 3 that appears as a superscript as an instruction to shift the decimal point three places when converting the number before the 10 into an ordinary decimal number. Often there will not be enough places available, but this is easily solved by adding some extra zeroes at the right. So in the example we begin by writing 3.11 as 3.110000, then move the decimal point three places to the right, giving 3110.00, and drop the unwanted zeroes to give 3110.

If the superscript is negative, for example if we have to convert 2.731×10^{-2} to a decimal number, we just interpret the minus sign as an instruction to shift the decimal place to the left instead of right, so, again writing some extra zeroes to make sure there will be a space, we first write 0002.731, move the decimal point two places to the left to give 00.02731, and then drop the unwanted zero at the beginning, giving 0.02731 as the result.

To go in the opposite direction, i.e. to write a decimal number such as 849.7 with an appropriate power of 10, we ask how far and in which direction the decimal point must be shifted to give a number before the power of 10 that is in the range 1–10? In this case the answer is two places to the right, so 849.7 becomes 8.497×10^2.

Sometimes the aim is not to get a number before the power of 10 that is in the range 1–10 but to have a power of 10 that is a multiple of 3, because many of the common prefixes for units represent such powers: m (milli-) means $\times 10^{-3}$, μ (micro-) means $\times 10^{-6}$, k (kilo-) means $\times 10^3$, etc. The rule for converting a decimal number is essentially the same except that we need to finish with a superscript of 3, −3, −6, etc. For example, 849.7 would become 0.8497×10^3, and if the number has units it is usual to incorporate the power of 10 into the units rather than writing it separately: 849.7 g is the same as 0.8497×10^3 g, or 0.8497 kg.

An obvious question raised by the last paragraph is why 3? What is special about powers of 10^3? The answer is that most of the quantities we need to measure in biochemistry are related to mass, which is proportional to volume for any substance of definite density, and as we live in a three-dimensional world this means length to the power 3. If we were working in a discipline more concerned with areas than with volumes, such as agronomy, we might well find it more convenient to have powers of 10 that were multiples of 2.

If we need to multiply two powers of 10 together the result is just the same as if we added the powers, and division is the same as subtracting powers: for example, $10^2 \times 10^3 = 10^{2+3} = 10^5$ and $\frac{10^6}{10^4} = 10^{6-4} = 10^2$. These are just specific instances of a general relationship that will be examined in Section 3.3. Conversely, if we want to add or subtract two numbers expressed as powers of 10, such as $6 \times 10^3 + 4 \times 10^2$ we must first relate them to the *same* power of 10, in this case writing them as $6 \times 10^3 + 0.4 \times 10^3$; then we can just add the two factors, to get $(6 + 0.4) \times 10^3 = 6.4 \times 10^3$.

1.6 **Operations that leave expressions unchanged**

Certain mathematical operations leave the expressions they act on unchanged. For example, we can multiply or divide any number by 1 and its value is unchanged; we can add 0 to or subtract 0 from any number and its value is unchanged. These four relationships may be written for an arbitrary value a as follows:

$$1 \times a = a, \qquad \frac{a}{1} = a$$

$$a + 0 = a, \qquad a - 0 = a$$

In these expressions a does not have to represent a number; it can represent any mathematical expression, so these relationships apply perfectly well to fractions. Moreover, '1' and '0' represent not only the numbers 1 and 0 but any mathematical expressions that have the values 1 and 0, respectively. Now any fraction in which the numerator and denominator are the same has the value 1, and so it follows that we can multiply any fraction by another fraction that has the same expressions for numerator and denominator without affecting its value. It is this general truth that allows us to cross-multiply when adding or subtracting fractions. All we are doing is multiplying each or any fraction in an expression by 1, and thus leaving it unchanged in value. In other words, we can multiply or divide *both* the numerator and denominator of any fraction by the *same* number or expression without affecting its value.

This should not be confused with the operations we can carry out on equations, which are more general. If two things are the same, then they remain the same if we change both of them in the same way. This means that with any equation we can apply almost any operation we like (with two exceptions as noted later in this section) to the left-hand side provided only that we carry out the same operation to the right-hand side. For addition and subtraction, this procedure is often described as 'moving a value from one side of an equation to the other and changing the sign'. This is acceptable as a shorthand way of describing what we do, but it should be avoided if it leads to any confusion about why it is legitimate. In the following example,

$$x - y = 3$$

we can move $-y$ from the left-hand side to the right-hand side, changing its sign so that it becomes $+y$:

$$x = 3 + y$$

and this is perfectly legitimate as long as one remembers that what one is 'really' doing is carrying out the same operation—adding $+y$—on both sides of the equation:

$$x - y + y = 3 + y$$

followed by adding $-y$ to $+y$ on the left-hand side to obtain nothing:

$$x = 3 + y$$

This may seem a rather long-winded way of achieving the same result. However, one should never be frightened of being long-winded in elementary mathematics: it is always better to be long-winded, but clear about what one is doing and why it is acceptable, than to use a quicker method that is only half-understood.

The similarity between how we handle equations and how we handle fractions is only superficial. We are allowed to multiply numerator and denominator of a fraction by the same number because this is equivalent to multiplying the whole fraction by 1, *not* because of any general permission to carry out equivalent operations on numerator and denominator. In fact multiplying or dividing by the same number is the *only* operation we can safely make on a fraction without changing its value. Thus,

$$\frac{x+3}{y+3} \text{ is not the same as } \frac{x}{y}$$

$$\frac{x^2}{y^2} \text{ is not the same as } \frac{x}{y}$$

$$\frac{\ln x}{\ln y} \text{ is not the same as } \frac{x}{y}$$

and so on. If you are not convinced by this, try each of these last three expressions with numerical values for x and y, such as $x = 3, y = 2$. In the first case, for example, $\frac{3+3}{2+3} = \frac{6}{5} = 1.2$, which is different from $\frac{3}{2} = 1.5$.

There is one important exception to the rule that we can carry out any operation on both sides of an equation, and to the rule that we can always multiply the numerator and denominator of a fraction by the same number. We should always add the qualification 'other than 0'. This may seem a little strange. After all, it is certainly true that if $x=y$ then $0 \times x = 0 \times y$ and so $0=0$, so if it is true then why are we not allowed to transform the equation in this way? The problem is that $0 \times x = 0 \times y$ and $0=0$ are *general truths*, whereas $x=y$ is a *particular truth*: true for the particular values of x and y that we are discussing, but not true in general. It is very convenient in mathematics to be able to apply an argument reversibly: not only do we want to proceed from a question to an answer, but we may also need to proceed from an answer to a question. For this to be possible we must be careful never to convert a specific statement into a general truth, because once we do this we lose any specific information that may have existed in the initial statement. The simplest way in which this can happen is from multiplying both sides of an equation by 0, and so we make a general prohibition against this operation. It is fairly easy to avoid as long as we are dealing with the explicit number 0, but the prohibition applies not only to the explicit number but also to expressions that are equal to zero: not only must we not multiply both sides of an equation by 0 but we must also, for example, avoid multiplying both sides by $x+5$ if x happens to have a value of -5. As it is not always obvious when one is violating the latter type of prohibition one should be cautious about multiplying by an algebraic expression unless it is certain that it cannot be zero.

In the case of fractions the prohibition takes the form that we must not multiply numerator and denominator of a fraction by 0. The question of whether any meaning can be attached to the fraction $\frac{0}{0}$ will not be discussed in this book. Suffice it to say that such a fraction is said to be *indefinite* and we should avoid introducing such fractions.

Even more generally, i.e. beyond the specific cases of equations and fractions, we should never divide anything by zero. As we have just seen, dividing zero by zero produces an indefinite result; dividing a non-zero quantity by zero is also sometimes said to give an indefinite result (in the sense that it is a result that is not defined in elementary mathematics), but the more common term in this case is *infinity*. Advanced mathematics has tools for comparing and classifying different kinds of infinities, but elementary mathematics does not, and it is best just to regard any operation that produces an indefinite or an infinite result as illegal. This is readily illustrated by any calculator, which will generate an error if you attempt to divide anything by zero.

1.7 **Simplifying fractions**

The fractions that result from a calculation are often more complicated than they need to be, in the sense that they are written in terms of numbers that are larger than necessary. The fraction $\frac{8}{24}$, for example, has the same value as $\frac{1}{3}$, and unless there is some special reason for writing it as $\frac{8}{24}$ it is nearly always better to write it as $\frac{1}{3}$. There are two points to examine here: how do we know that $\frac{8}{24}$ is the same as $\frac{1}{3}$, and, what general procedure will allow us to find the simplest way of writing a fraction?

First, we know that $\frac{8}{24}$ is the same as $\frac{1}{3}$ because $\frac{8}{24}$ is what we get if we multiply both the numerator and the denominator of $\frac{1}{3}$ by 8, and, as discussed in the previous section, multiplying the numerator and denominator of a fraction by the same number (other than 0) is equivalent to multiplying the whole fraction by 1 and has no effect on its value. The reverse operation, dividing both the numerator and the denominator of $\frac{8}{24}$ by 8, is likewise equivalent to dividing by 1, and has no effect on the value.

The general procedure for simplifying a fraction expressed in numbers is to express the numerator and denominator in terms of their *prime factors*, i.e. in terms of numbers that cannot be obtained by multiplying smaller numbers together. Once we do this we can recognize at sight which factors are common to the numerator and the denominator, and we can then cancel them just as we cancelled 8 in the example of $\frac{8}{24}$.

To find the prime factors of a number it is useful to have a method for recognizing which small numbers will evenly divide a particular large number. Fortunately, there are simple rules for recognizing this for all numbers from 1 to 12 apart from 7. In all of these rules, which are listed in **Box 1.3**, the word 'if' really means 'if and only if', i.e. the conditions given are necessary as well as sufficient. (Mathematicians sometimes write 'if' in this sense as 'iff'.) It may seem that the rules are rather arbitrary and that there are too many of them to remember. However, the task can be simplified by realizing that several of the numbers in the list are *composite*, i.e. they have factors: $4 = 2 \times 2$; $6 = 2 \times 3$; $8 = 2 \times 2 \times 2$; $9 = 3 \times 3$; $10 = 2 \times 5$; $12 = 2 \times 2 \times 3$. We can ignore these, because if a number is not divisible by 3, for example, it is certainly not divisible by 6, 9 or 12. This leaves just 2, 3, 5, 7 and 11 to be tested. The rules for 2 and 5 are both simple and familiar, so they should cause no trouble, and there is no rule for 7, so we are left with just 3 and 11, which we can now illustrate with some examples.

The digits of the number 187 add up to $1 + 8 + 7 = 16$, which is not divisible by 3, so 187 is not divisible by 3. Giving the digits alternate + and − signs, we have $1 - 8 + 7 = 0$, which is divisible by 11 (0 is considered to be divisible by any number apart from 0, an important point to remember in applying the rule for 11), and so 187 is divisible by 11. As another example, 348 is divisible by 3, because $3 + 4 + 8 = 15$ is divisible by 3, but it is not divisible by 11, because $3 - 4 + 8 = 7$ is not divisible by 11.

These rules can be applied to numbers of any size, but for large numbers we may often need to apply them more than once to reach a conclusion. For

Box 1.3 **Determining the factors of a number**

Rule	*Examples*
1 All integers are evenly divisible by 1.	82, 239
2 An integer is evenly divisible by 2 if its last digit is divisible by 2, i.e. its last digit is 0, 2, 4, 6 or 8.	20, 46, 182
3 An integer is evenly divisible by 3 if the sum of its digits is divisible by 3.	$36\ (3+6=9)$
4 An integer is evenly divisible by 4 if the number formed by its last two digits is divisible by 4.	$1084 = 1000 + 84$
5 An integer is evenly divisible by 5 if its last digit is divisible by 5, i.e. its last digit is 0 or 5.	. 40, 695
6 An integer is evenly divisible by 6 if it is divisible both by 2 and by 3.	42, 126
7 There is no rule for determining if a number is divisible by 7*.	—
8 An integer is evenly divisible by 8 if the number formed by its last three digits is divisible by 8.	$95\,128 = 95\,000 + 128$
9 An integer is evenly divisible by 9 if the sum of its digits is divisible by 9.	$738\ (7+3+8=18)$
10 An integer is evenly divisible by 10 if its last digit is divisible by 10, i.e. its last digit is 0.	430, 8190
11 An integer is evenly divisible by 11 if the sum obtained by alternately adding and subtracting its digits is divisible by 11.	$1067\ (1-0+6-7=0)$
12 An integer is evenly divisible by 12 if it is divisible both by 3 and by 4.	84, 132

*However, there is no difficulty with the help of a calculator. To decide if 189 is divisible by 7, for example, do the division on a calculator. The result is 27, an integer, so 189 is divisible by 7. With 219 the result would be 31.2857..., which is not an integer, so 219 is not divisible by 7.

example, the digits of 372 915 293 517 add up to 54, and if you do not recognize immediately if this is divisible by 3 you can repeat the operation, and note that $5+4=9$ is divisible by 3 (and also by 9), so 372 915 293 517 is divisible by 3 (and also by 9). Giving the digits alternate $+$ and $-$ signs they add up to -30, and $-3+0=-3$ is not divisible by 11, so 372 915 293 517 is not divisible by 11 either.

Suppose now we have a fraction $\dfrac{234}{1365}$ that we need to simplify. For both numerator and denominator, much the easiest procedure is to apply the easiest rules first. By rule 2, 234 is divisible by 2, so it can be written as 2×117. 117 is not divisible by 2, but rule 3 allows us to recognize that it is divisible by 3 and can be written as 3×39. 39 is also divisible by 3 and can be written as 3×13, so 117 is $3 \times 3 \times 13$, and 234 is $2 \times 3 \times 3 \times 13$. Notice that checking directly whether a number is divisible by 9 is less work than checking if it is divisible by 3, then dividing by 3 and checking if the result is divisible by 3: this is similar to the case of $10 = 2 \times 5$ discussed below.

Looking now at 1365, it is not divisible by 2, but rule 5 is easier to apply than rule 3, so we apply it first, and write 1365 as 5×273. 273 is divisible by 3, so it is 3×91. Here we are unlucky, because 91 is one of the only two numbers less than 100 that have factors but cannot be factorized at sight (the other being 49). None of the rules gives a positive result, so we are forced, as a last resort, to try dividing by 7, and 91 turns out to be 7×13.

It follows that the original fraction $\frac{234}{1365}$ can be simplified as follows:

$$\frac{234}{1365} = \frac{2 \times 3 \times 3 \times 13}{3 \times 5 \times 7 \times 13} = \frac{2 \times 3 \times \cancel{3} \times \cancel{13}}{\cancel{3} \times 5 \times 7 \times \cancel{13}} = \frac{2 \times 3}{5 \times 7} = \frac{6}{35}$$

It is perhaps worth pointing out also that although 10 is composite it is so easy to recognize whether a number is divisible by 10 (by the final 0) and to divide by 10 that it is always worthwhile extracting any factors of 2×5 first rather than treating 2 and 5 separately from the beginning. For example, although it would be perfectly correct to factorize 350 by extracting 2 first, followed by 5 and then 5 again:

$$350 = 2 \times 175 = 2 \times 5 \times 35 = 2 \times 5 \times 5 \times 7$$

it is easier and quicker to divide by 2×5 at the outset:

$$350 = 2 \times 5 \times 35 = 2 \times 5 \times 5 \times 7$$

1.8 **Probability as a mathematical concept**

In ordinary life we often use such words as 'probable', 'probability', 'likely' and 'likelihood'. In mathematics, these words have technical and precise meanings (so we should avoid using them loosely in a mathematical context). Likelihood is too advanced a topic for this book (especially as regards its differences from probability), but some understanding of probability is fundamental for understanding the basic ideas of statistics and their application to scientific questions.

The essential idea of a mathematical probability is that it is the reciprocal of the number of *equally probable* events that could occur rather than the one of interest. For an unbiassed dice each of the six faces has an equal probability of lying uppermost when the dice falls, so the probability of throwing a 6 is $\frac{1}{6}$. Two essential properties of probabilities calculated in this way are that they are *additive* for combinations of exclusive possibilities, and *multiplicative* for independent events considered as one event. What this means is that the probability of getting better than a 4 in one throw is $\frac{1}{6} + \frac{1}{6} = \frac{1}{3}$, and that if we throw two dice (or the same dice twice) then the probability of getting 6 in both throws is $\frac{1}{6} \times \frac{1}{6} = \frac{1}{36}$. The additive property is intuitively fairly obvious, but the multiplicative property may be less so. To understand where it comes from it is useful to make a list of all the possible ways in which two throws could go, using a notation where 1,5 means a 1 in the first throw followed by a 5 in the second (etc.):

$$
\begin{array}{cccccc}
1,1 & \mathbf{1,2} & 1,3 & \mathbf{1,4} & 1,5 & \mathbf{1,6} \\
2,1 & \mathbf{2,2} & 2,3 & \mathbf{2,4} & 2,5 & \mathbf{2,6} \\
3,1 & \mathbf{3,2} & 3,3 & \mathbf{3,4} & 3,5 & \mathbf{3,6} \\
4,1 & \mathbf{4,2} & 4,3 & \mathbf{4,4} & 4,5 & \mathbf{4,6} \\
\mathit{5,1} & \mathbf{\mathit{5,2}} & \mathit{5,3} & \mathbf{\mathit{5,4}} & \mathit{5,5} & \mathbf{\mathit{5,6}} \\
\mathit{6,1} & \mathbf{\mathit{6,2}} & \mathit{6,3} & \mathbf{\mathit{6,4}} & \mathit{6,5} & \mathbf{\mathit{6,6}}
\end{array}
$$

(Ignore the italic and bold type for the moment.) If you count these, you will see that there are 36 different outcomes possible, each of which is equally probable if the dice is unbiassed, but only one of them (the last in the list) shows a 6 in both throws. The property applies to more complicated outcomes that we might want to study. For example, the probability of getting better than a 4 in the first throw is $\frac{1}{6} + \frac{1}{6} = \frac{1}{3}$, as stated already, and the probability of getting an even number (2, 4 or 6) in the second is $\frac{1}{6} + \frac{1}{6} + \frac{1}{6} = \frac{1}{2}$, so the probability of getting better than a 4 in the first throw *and* an even number in the second is $\frac{1}{3} \times \frac{1}{2} = \frac{1}{6}$. If we count all of the possibilities in the table that satisfy this (made easy by the fact that all the entries showing better than 4 in the first throw are shown in *italics*, whereas all the entries showing an even number in the second are shown in **bold**), we can readily confirm that there are just the six that are shown in both italic and bold, and as $\frac{6}{36}$ is the same as $\frac{1}{6}$ this agrees with the result obtained by multiplying $\frac{1}{3}$ by $\frac{1}{2}$.

This is fine as far as it goes, but as soon as we want to apply these ideas to a real scientific question we run into two difficulties. The first is that in the real world we are rarely dealing with equally probable events, and we may not know what the actual probabilities are, so that we are in the position of the gambler who knows that the dice produced by his opponent is biassed but who does not know the details of the bias. The second is that the total number of outcomes to be compared is often enormously too large to count. So, if we are interested in knowing how likely it is that the word CAT (where C, A and T are three of the four bases found in DNA) occurs at least once in a DNA sequence of 100 bases, we first have to face the fact that the four bases are not equally frequent, and second that the number of different sequences of 100 bases is $4 \times 4 \times 4 \times 4 \times \cdots \times 4$, with 100 4's multiplied together, i.e. about 16 followed by 59 zeroes. So, obviously it is not practical to list all the possibilities and count how often the word CAT occurs. We therefore need to come back to this question in a more serious way, and will do so in Chapter 8. Here we shall continue this elementary discussion by examining that difference between *permutations* and *combinations*.

1.9 **Permutations and combinations**

Both permutations and combinations are concerned with combining groups of events in such a way that any particular kind of event can only occur once, i.e. once it has been chosen it cannot be chosen again. They thus have more in common with hands of bridge than with tossing dice: if you toss a 6 in one throw you are still able to throw 6 in the second and third as well, but in a hand of bridge you can have the ace of spades but you cannot have two aces of spades.

The essential difference between permutations and combinations is that in studying combinations we do not care about the order in which the different events occur whereas in studying permutations we do. The value of a bridge hand containing the ace of spades and the ace of diamonds does not depend on which of the two aces was dealt first, so this is a matter of combinations, but when getting dressed in the morning the result of putting on your underpants before your trousers is quite different from the result of putting on the same two items in the opposite sequence, so this is a matter of permutations. (To confuse matters, the word *perm* as used in the football pool industry refers to combinations of different outcomes in football matches, not permutations; the same term used by a hair stylist has nothing to do with either, and nor does the Russian city of Perm.)

Now let us consider these terms in relation to a particular example. The first four reactions of glycolysis involve four kinds of reactions, 6-phosphorylation (glucose to glucose 6-phosphate, catalysed by hexokinase), isomerization (glucose 6-phosphate to fructose 6-phosphate, catalysed by hexose phosphate isomerase), 1-phosphorylation (fructose 6-phosphate to fructose 1,6-bisphosphate, catalysed by phosphofructokinase) and disproportionation (fructose 1,6-bisphosphate to dihydroxyacetone phosphate and glyceraldehyde 3-phosphate, catalysed by aldolase). We could conceive of the same final result being achieved by following the same four steps in a different order, but having once done a 6-phosphorylation we cannot do it a second time as there is only one 6-position to phosphorylate. If we were interested in analysing different ways in which glycolysis might have evolved, to decide, for example, if the version of glycolysis that actually occurs has identifiable chemical advantages over others, then we should need to examine such questions as the identity and chemical reactivity of the metabolite produced after two steps, or energetic obstacles that would need to be surmounted during the first three steps, etc.

If we code the four possible steps by **6, i, 1** and **d** (listed in the order they occur in real glycolysis), we can code the real process as **6i1d**. Then, for the first two steps we have the following possibilities: **6i, 61, 6d, i6, i1, id, 16, 1i, 1d, d6, di** and **d1**. (You may object that **d6** is impossible because neither glyceraldehyde nor dihydroxyacetone has a 6-position, but the difficulty disappears if we use the code 6 to mean phosphorylation of the group that was originally the 6-position of glucose.) Now, if we are just interested in the product after two steps, then **6i** and **i6** are equivalent because they both produce fructose 6-phosphate; likewise **61** and **16, 6d** and **d6, i1** and **1i, 1d** and **d1,** and **id** and **di** are, respectively, equivalent. So the 12 *permutations* of reactions for the first two steps produce six non-equivalent *combinations*, with six different possible products. But if we are interested in the chemical feasibilities of the different possibilities then we need to consider not only the final result but the identity of the intermediate, and then we must study all 12 permutations.

Notice that the mathematician has no way of knowing which of these is 'right': it depends on what chemical questions you want to ask, and so depending on the circumstances it is sometimes necessary to calculate the number of permutations of n objects taken m at a time, which can be symbolized as ^nP_m, and sometimes the number of combinations, symbolized as ^nC_m.

The number of permutations could have been calculated instead of counted by reflecting that there are four ways of choosing the first step, but once this is chosen only three remain for choosing the second. So for two steps there are $4 \times 3 = 12$ permutations. Each of these contains two reaction types, and so if we do not care about the order we must have $\dfrac{12}{2} = 6$ combinations.

More generally, we can calculate the number of permutations of n events considered m at a time as $n(n-1)(n-2)\cdots$, with m terms in the multiplication, i.e. stopping at $(n-m+1)$. As this is rather a messy way of expressing the relation (though it describes the easiest way of calculating it), it is more usually written as follows:

$$^nP_m = n(n-1)(n-2)\cdots(n-m+1)$$

$$= \frac{n(n-1)(n-2)\cdots 3 \times 2 \times 1}{(n-m)(n-m-1)(n-m-2)\cdots 3 \times 2 \times 1}$$

$$= \frac{n!}{(n-m)!}$$

For the example of the possible permutations of glycolytic reactions, this formula shows that there are $\dfrac{4!}{(4-3)!} = \dfrac{4!}{1!} = 24$ ways of arranging the first three steps (**6i1, 6id, 61i, 61d, 6di, 6d1, i61, i6d, i16, i1d, id6, id1, 16i, 16d, 1i6, 1id, 1d6, 1di, d6i, d61, di6, di1, d16** and **d1i,** with the same coding as above).

Stopping the first multiplication at $n-m+1$ is just a way of ensuring that there are m terms, but we can ensure the same thing by continuing the multiplication all the way to 1, as in the numerator of the middle expression, but then dividing by the result of starting the same multiplication at $n-m$. Finally $n!$ is just a shorthand way of writing $n(n-1)(n-2)\cdots 3 \times 2 \times 1$, so that the right-hand fraction is exactly the same as the middle one; it is just a more compact way of writing the same thing.

Notice that numerical expressions of this kind can easily be evaluated with a calculator, even a simple one that has no specific function key for it. For example, $5! = 5 \times 4 \times 3 \times 2 \times 1 = 120$. Calculators with a specific key (normally labelled !) just do the same thing in one step. Notice also, however, that $n!$ gets large extremely fast as n becomes larger:

$$0! = 1, \quad 1! = 1, \quad 2! = 2, \quad 3! = 6, \quad 4! = 24, \quad 5! = 120$$

$$6! = 720, \quad 7! = 5040, \quad 8! = 40\,320, \quad 9! = 362\,880, \quad 10! = 3\,628\,800$$

so the simpler calculators are likely to give errors for quite small values of n.

The ! sign is called a *factorial sign*, and speaking out loud one would pronounce $n!$ as 'n factorial'. Mathematical writing does not indulge in a great many exclamations (in part because any mathematical result that is properly understood cannot be surprising), so there is little danger of confusing a factorial sign with an exclamation point. However, writers who like to label their more remarkable statements so that they will not pass unnoticed need to take care not to put exclamation points in places where they could be read as factorial signs. 'I was expecting to find 6 enzymes in the pathway, but actually there are 4!'—is this surprisingly few or surprisingly many?

If we consider the number of permutations of n events taken n at a time (i.e. putting $m = n$) then the first formula for ^nP_m above remains clear enough and must have the value $n(n-1)(n-2)\cdots 3 \times 2 \times 1$, or $n!$, but the second seems to make no sense at all, and the third gives $\frac{n!}{0!}$, which looks quite bizarre. The problem disappears if we *define* $0!$ as 1 and do not try to understand why it should have this value. It is clearly the only way to define $0!$ so that the third line is to mean the same thing as the first line, and this proves to be true in general: if we interpret $0!$ as 1 whenever we encounter it, it will always give a result that agrees with the result of simple counting.

As this result seems so peculiar I should perhaps try to rationalize it a little more. If we count backwards, it is obvious from inspection that we can calculate 4! from 5! by dividing by 5, we can calculate 3! from 4! by dividing by 4, we can calculate 2! from 3! by dividing by 3, and we can calculate 1! from 2! by dividing by 2. We should not be surprised, therefore, if dividing 1! by 1 gives 0!, in other words that $0! = 1$.

Resuming, it follows that $n!$ is in general the number of sequences in which n objects, events, etc. can be arranged.

This now allows us to come back to the question of the number of combinations of n objects taken m at a time. We have already seen that the number of permutations

Table 1.1 Permutations and combinations.

n	$\dfrac{n!}{(n-m)!}$ permutations for $m=\cdots$							$\dfrac{n!}{m!(n-m)!}$ combinations for $m=\cdots$						
	0	1	2	3	4	5	6	0	1	2	3	4	5	6
1	1	1						1	1					
2	1	2	2					1	2	1				
3	1	3	6	6				1	3	3	1			
4	1	4	12	24	24			1	4	6	4	1		
5	1	5	20	60	120	120		1	5	10	10	5	1	
6	1	6	30	120	360	720	720	1	6	15	20	15	6	1

is $\dfrac{n!}{(n-m)!}$, but if we do not care about the order we note that each permutation is one of $m!$ ways of arranging the same m objects in sequence, and thus the number of combinations can be obtained by dividing the number of permutations by $m!$, and the formula is thus as follows:

$$^{n}C_{m} = \frac{n!}{m!(n-m)!}$$

It is useful to get an idea of what these formulas mean in practice by examining some numerical examples, which may be found in Table 1.1. If m is zero, then the numbers of permutations and combinations are both 1, regardless of the value of n: there is only one way of doing nothing. If $m=1$ then the numbers of permutations and combinations are equal to one another and to n: there are n ways of making one choice out of n possibilities, and if there is only one choice the question of order does not arise. However, if m is greater than 1 then the number of permutations will always be larger than the number of combinations, and may be very much larger $^{6}P_{5}=720$, whereas $^{6}C_{5}=6$. Notice finally that although the number of permutations increases as m increases and never decreases, the number of combinations increases at first and then decreases back to 1 when $m=n$. This behaviour can be rationalized by reflecting that taking m objects out of a bag containing n objects is equivalent to deciding which $(n-m)$ to leave in the bag, and thus involves the same number of choices.

1.10 Having a rough idea of the answer

An additional reason why we do not find calculations involving small amounts of money a problem is that when handling small financial transactions we always have a rough idea of what the answer ought to be without doing any calculation at all, and we know that if there is an error it is unlikely to bankrupt us. In scientific calculations it is very useful to cultivate the habit of thinking about what sort of magnitude the answer is likely to have: is 20 pmol (20 picomoles) of a hormone a large or a small amount? Should we be surprised or not to be told that two proteins bind together with a dissociation constant of 3 mM? Anyone doing a lot of scientific calculations needs to have a rough idea of the answer to these and many similar questions.

In the financial transaction described in Section 1.3, we might not know that we expected exactly 115 cents in change, but we should certainly know at the outset that the change would be more than 10 cents and less than 2 dollars, and that, whatever amount it was, we would not be bankrupted by a mistake. The fact that we do not have this built-in knowledge about how much the money is worth is one reason why it is so easy to be cheated when one is away from home: is 150 000 dongs a good price for a bowl of rice in Vietnam? Would 3000 lek be enough to buy a house in Albania? Who knows?*

The moral of this from the point of view of scientific calculations is that we should always try to express results in terms of small numbers and, where possible, in terms of familiar units: if an association constant turns out to be about 1 500 000 M^{-1} it is not immediately easy to know if this is a reasonable value or not, first because 1 500 000 is too large a number to be easily visualized, and second because we do not think often enough about reciprocal concentrations to know what sort of ranges to expect. We can overcome the second problem by calculating the corresponding dissociation constant as the reciprocal, 0.000 000 667 M; most of us find concentrations easier to think about than reciprocal concentrations, but the number is now too small to visualize easily, so we convert to more appropriate units, as 0.667 μM. If this comes from an experiment in which we are studying the association of two proteins at concentrations in the micromolar range then it is easy to see that it is a reasonable sort of value. If the authors referred to at the end of Section 1.1 had thought in these terms they would have been unlikely to have reported an association constant of 0.000 003 M^{-1} for the mutual binding of two proteins.

A complete list of all the magnitudes of common units used in biochemistry would be too voluminous to justify inclusion here, but a few examples will illustrate the sort of thinking that one should cultivate to avoid making absurd mistakes. Let us start with molecular mass, as perhaps the most fundamental property for a chemist. We all know that a hydrogen atom weighs 1 Da (dalton), and that all of the natural elements have atomic masses less than 100 Da. All of the common components of biochemical molecules (C, H, O, N, S, P) have atomic masses less than 20, and this tells us immediately that the common metabolites like glucose, ATP, NAD, glycine, etc., with 10–50 atoms per molecule, will typically have molecular masses in the range 50–500 Da, so a calculation purporting to show a metabolite of supposedly average size with a molecular mass of 15 Da has clearly given a result that is too small; conversely, 5000 Da would clearly be too big

---

* If you really want to know, you can get an idea from the fact that in January 1998 there were about 20 000 dongs to the pound, or about 12 000 to the dollar. So 150 000 dongs is about £7.50, on the expensive side for a bowl of rice. In the same period there were about 250 lek to the pound, so 3000 lek, or about £12, would be inexpensive for a house, even in Albania.

for anything other than a polymer. As we know that glycine is the lightest of the amino acids we would expect its molecular mass of 75 Da to be somewhat smaller than average, so 100 Da/residue is a reasonable guess for the average molecular mass of a protein. An estimate of 105 Da/residue is more accurate, but 100 Da/residue is much easier to calculate with and is good enough for the sort of mental calculation we are discussing here; much better, in fact—because the aim is to get a rough idea of the answer and for this it is important to avoid unnecessary complexity that may lead to a completely wrong answer.

Armed with this information, we can expect a protein of 50 000 Da to have around 500 residues, and conversely we expect a protein of 350 residues to have a molecular mass of around 35 000 Da, and so on. These are typical values for many enzymes, but the numbers are already inconveniently big: it is better to convert them to 50 kDa and 35 kDa respectively, and to think of each kDa as corresponding to about ten residues (or nine if you want to be precise).

What sort of numbers correspond to these in the real world? A bag of sugar from the supermarket typically weighs 0.5 kg, and the molecular mass of sucrose $(C_{12}H_{22}O_{11})$ is 342, so 1 mol is 0.342 kg. However, hydrolysis yields two molecules of hexose per molecule of sucrose, and as hexoses occur much more often than sucrose in elementary biochemistry it is perhaps better to think of a bag of sugar from the supermarket as equivalent to around 3 mol of hexose, so if we dissolve all of it in water and make it up to 1 L we shall have a solution of about 3 mol/L, or 3 M, in hexose equivalents. This is quite a concentrated solution in terms of ordinary biochemistry, so we should see immediately that very few concentrations in biochemical experiments are likely to be much more than 1 M; 8 M urea and 8 M guanidine hydrochloride, used as denaturing agents, are almost the only exceptions. Concentrations in the mM range are much more typical for everyday metabolites like glucose, for which the concentration in the blood is typically about 5 mM. Carrying on this sort of investigation we can draw up a list of the sort of things that are likely to have concentrations in particular ranges, as is done in **Box 1.4**. The different ways that are used in biochemistry for writing volumes and concentrations are considered in **Box 1.5**.

Physical constants that have units of concentration, like dissociation constants, Michaelis constants, inhibition constants, etc., can usually be defined so that they correspond, exactly or at least approximately, to real concentrations of real things. A Michaelis constant, for example, is the concentration of substrate at which the rate is half of the limiting rate, which means that varying the substrate concentration in a range near the K_m will typically produce an easily observable variation in rate. Thus, a K_m of 1 nM for a common metabolite is too small to be measured, whereas a K_m of 5 M is too large: if they come out like this in a calculation, there is certainly a mistake.

You can apply this sort of analysis to all the kinds of quantities that are measured in biochemistry: for example, wavelengths of radiation, volumes of cells, forces developed by muscles, muscle fibres, and actin filaments, and so forth. The

---

Box 1.4 **Typical concentrations in biochemistry**

Around 1 nM	Enzymes, other proteins, drugs
Less than 1 μM	Trace metabolites
1 μM–0.1 mM	More labile metabolites (fructose 1,6-bisphosphate...)
0.1–5 mM	Typical metabolites in the cell (ATP, NAD, glucose...)
0.1–1 M	Buffer salts (phosphate...)
1 M	Artificially concentrated solutions of sugars
More than 5 M	Nothing apart from urea, guanidine HCl and water
More than 50 M	Water (pure water being 55.6 M)

---

principle is always the same: that it is desirable to have a rough idea of the likely magnitudes of the quantities one has to deal with, and to express measured constants in units as familiar as possible. It is not necessary, therefore, to continue the discussion in more detail. To conclude, it may be helpful to reflect on an everyday example far removed from practical biochemistry. The fuel consumption of a car is typically of the order of 6 L per 100 km. As this is a volume divided by a length it is an area: express it in more usual units for an area, and try to attach a physical meaning to the area that it measures.

Box 1.5 **Writing volumes and concentrations**

Strictly speaking the SI unit of volume is the cubic metre, or m^3, but this is far too large for ordinary use in chemistry and is always replaced by the cubic decimetre, or dm^3 ($1000\,dm^3 = 1\,m^3$). Neither this name nor the corresponding symbol are commonly used in chemistry and biochemistry, however, being replaced by the litre, which is exactly equivalent to the cubic decimetre, and has alternative symbols l and L. Although the lower-case l is older and accords better with the usual practice of restricting the use of capital letters to symbols for units named after people (such as Da for dalton), it is easily confused with the number 1, and the capital L is also permitted and is becoming quite common, especially if it stands alone, i.e. many people who would rarely write 5 mL will still often write 5 L, because 5 ml is clear as it stands whereas 5 l could be read as a number.

Concentrations are usually expressed in moles per litre, i.e. $mol\,L^{-1}$, but this is often regarded as a unit in its own right, with the name 'molar' and symbol M, so $1\,mol\,L^{-1}$ is exactly the same as 1 M. Both ways of writing concentrations are common in biochemistry and both are used in this book. It is important to distinguish between 1 mol, a quantity, and 1 M, a concentration, and *never* to use mol as an abbreviation or symbol for molar.

1.11 Appendix: Notes and solutions to the problems in Box 1.1

1. $\frac{14}{35} = \frac{7 \times 2}{7 \times 5} = \frac{2}{5}$, because $14 = 7 \times 2$ and $35 = 7 \times 5$, i.e. the numerator and denominator have a common factor 7 that can be cancelled.

2. $\frac{2}{3} + \frac{3}{5} = \frac{2 \times 5}{3 \times 5} + \frac{3 \times 3}{5 \times 3} = \frac{10}{15} + \frac{9}{15} = \frac{19}{15}$. The result can also be written as a 'mixed fraction', i.e. as $1\frac{4}{15}$, because $19 = 15 + 4$ and $\frac{15}{15} = 1$, but this this type of symbol is little used in scientific applications.

3. $\frac{2}{3} \times \frac{3}{5} = \frac{2 \times 3}{3 \times 5} = \frac{2}{5}$ (cancelling the common factor of 3).

4. The value is the same as in question 3, i.e. $\frac{2}{5}$, but to express it as a decimal the denominator 5 must be replaced by 10, which is done by multiplying both numerator and denominator by 2: $\frac{2}{5} = \frac{2 \times 2}{2 \times 5} = \frac{4}{10} = 0.4$. If the two fractions are converted to decimals before multiplying them together the result would be $0.667 \times 0.6 = 0.4002$. The different result is explained by the fact that 0.667 is an approximate not an exact representation of $\frac{2}{3}$.

5. $0.75 = \frac{75}{100} = \frac{3 \times 5 \times 5}{2 \times 2 \times 5 \times 5} = \frac{3}{2 \times 2} = \frac{3}{4}$.

6. Write 8.37 with some extra zeroes at the left: 00008.37. Then move the decimal point three places to the left: 00.00837. Then delete the superfluous zero: 0.00837.

7. $(x+y)(y-z) = (3+2)(2-5) = 5 \times (-3) = -15$.

8. $2x + 7 = 13$, so $2x = 13 - 7 = 6$, so $x = 6/2 = 3$.

9. $3^3 = 3 \times 3 \times 3 = 27$.

10. Substituting $x = 4$ into $7x + 3 = 3y + 1$ gives $7 \times 4 + 3 = 3y + 1$, hence $28 + 3 = 3y + 1$, so $3y = 28 + 3 - 1 = 30$ and $y = \frac{30}{3} = 10$.

11. If the surface area is A, the radius is r and the circumference at the equator is c, then the relationship expressed in words in the question may be written algebraically as $A = 2rc$. Substituting $c = 2\pi r$ into this gives $A = 2r \times 2\pi r = 4\pi r^2$.

12. 15 is not a prime because it can be expressed in terms of prime factors as $15 = 3 \times 5$. None of the other numbers has any prime factors other than itself and 1.

13. $42 = 2 \times 21 = 2 \times 3 \times 7$.

14. $(a+b)(a-b) = a^2 - ab + ab - b^2 = a^2 - b^2$.

15. $(x^2 - y^2) = (x+y)(x-y)$. Apart from the change in symbols this is the same as question 14 in reverse.

16. $4! = 4 \times 3 \times 2 \times 1 = 24$.

17. $3.8\,\text{mM} = 3.8 \times 10^{-3}\,\text{M}$, so its reciprocal is $\frac{1}{3.8} \times 10^3 \text{M}^{-1} = 0.263 \times 10^3\,\text{M}^{-1} =$ $263\,\text{M}^{-1}$. To be considered correct, *both* the number (263) *and* the unit (M^{-1}) must be correct: giving the answer as $263\,\text{M}$ should no more be considered adequate than giving it as $0.263\,\text{M}^{-1}$ would be.

18. There are four ways of deciding which letter to put first, and once this is decided there are only three ways of choosing which to put second. Once the first two are decided there are only two ways of choosing the third, and once this is decided only one possibility remains for the last. Thus, the total number of ways is $4 \times 3 \times 2 \times 1 = 24$. See also the *Note* at the end of the answer to the next question.

19. When the question is worded in this way there is no prohibition against using any letter more than once (so a sequence like AAAA is allowed whereas it would not have been allowed in the answer to question 18). Thus, in this problem there are four possible choices for each of the four positions, giving a total of $4 \times 4 \times 4 \times 4 = 256$ possible sequences altogether.

Note. There may be some grounds for considering that the wording of questions 18 and 19 was insufficiently precise to distinguish clearly between the intended meanings. The different meanings do correspond to the ways in which the different wordings are often used in biochemistry, but they could be made clearer by adding a parenthesis to question 18 stating that each letter was to be used exactly once, and a different parenthesis to question 19 stating that there was no restriction on the number of times each letter could be used.

20. There are four ways of choosing the first, and once this is chosen there remain three ways of choosing the second, so altogether there are $4 \times 3 = 12$ possibilities (AC, AG, AT, CA, CG, CT, GA, GC, GT, TA, TC, TG)? However, as each pair occurs twice in this list (CA as well as AC, etc.), this result must be divided by 2 to take account of the fact that order is not important, so the result is $\frac{12}{2} = 6$.

21. This is the number already calculated as an intermediate stage in question 20, i.e. $4 \times 3 = 12$.

1.12 **Problems**

1.1 Remove the brackets from the following expressions and combine like terms:

(a) $x + 3y - z + 2xy + 3z - y + 2x$

(b) $2 - 3\,(x+y)$

(c) $5(x - 2y) - 3(y + 3x)$

(d) $(A - 3B) + (2A + 3C)$

(e) $(x + 4y) - (2x + y)$

(f) $2(4x + 4y) + (2y - x)$

(g) $(3x + y) - (2y - z)$

(h) $(3x+y)-(2y-z)+3(2x+y)$

(i) $(4a-3b+c)+(a+2b+3d)$

(j) $(4p-3q+r)-(q-5p+3r)$

(k) $3x^2-[4x+2x(y-x)]$

1.2 Remove the brackets from the following expressions and combine like terms:

(a) $(A-3B)(2A+3C)$

(b) $(x+4y)(2x+y)$

(c) $2(4x+4y)(2y-x)$

(d) $(3x+y)(2y-z)$

(e) $(4a-3b+c)(a+2b+3d)$

(f) $(4p-3q+r)(q-5p+3r)$

(g) $(4+x)(4-x)$

(h) $(2+x)(2+x)$

1.3 Factorize the following expressions:

(a) $4x+2y-6z,$ \qquad (b) $p^2+2pq+q^2$

(c) $p^2-2pq+q^2,$ \qquad (d) $p^2-4pq+4q^2$

(e) $9x^2+6xy+4y^2,$ \qquad (f) A^2-B^2

(g) $4u^2-9v^2$

1.4 Which of the following fractions can be expressed more simply, considering *obvious* factorizations only, i.e. numerical factors, differences between two squares and perfect squares:

(a) $\dfrac{2x+6y}{4y-8x}$, \quad (b) $\dfrac{a^2-4b^2}{a+2b}$, \quad (c) $\dfrac{p-3}{p+3}$

(d) $\dfrac{(u-v)^2}{u^2-v^2}$, \quad (e) $\dfrac{25-4x^2}{5-2x}$

1.5 Express each of the following fractions so that it has a denominator of 100:

(a) $\dfrac{1}{4}$, \quad (b) $\dfrac{3}{5}$, \quad (c) $\dfrac{7}{2}$

(d) $\dfrac{3}{10}$, \quad (e) $\dfrac{4}{25}$

1.6 Express each of the following fractions so that the numbers in numerator and denominator are as small as possible:

(a) $\dfrac{63}{75}$, \quad (b) $\dfrac{4}{8}$, \quad (c) $\dfrac{27}{12}$

(d) $\dfrac{3}{17}$, \quad (e) $\dfrac{15}{35}$

1.7 Evaluate each of the following expressions, leaving each answer as an integer (if possible) or otherwise as a fraction with the smallest possible numbers in numerator and denominator:

(a) $\dfrac{3}{7} + \dfrac{2}{5}$, (b) $\dfrac{3}{8} - \dfrac{1}{5}$, (c) $\dfrac{3}{8} + \dfrac{10}{16}$

(d) $\dfrac{5}{7} - \dfrac{6}{11}$, (e) $\dfrac{7}{6} - \dfrac{4}{15}$, (f) $\dfrac{1}{3} + \dfrac{4}{5} - \dfrac{3}{7}$

1.8 Express each of the following as a single fraction, without multiplying out any brackets:

(a) $\dfrac{x}{y+5} + \dfrac{y+1}{x}$, (b) $\dfrac{a+3}{a+5} + \dfrac{b+1}{b-2}$, (c) $\dfrac{Va}{K+a} + ka$

1.9 Write each of the following decimal numbers as a fraction, with the smallest possible numbers in the numerator and denominator:

(a) 8.35, (b) -32.1, (c) 0.75, (d) 1.375

1.10 Express each of the following fractions as a decimal number rounded to three decimal places:

(a) $\dfrac{4}{7}$, (b) $\dfrac{8}{13}$, (c) $\dfrac{22}{7}$, (d) $\dfrac{13}{20}$

1.11 Which of the answers to question 10 would be different if the result was truncated to three decimal places rather than rounded?

1.12 What would be the result of each of the following operations:
 (a) increasing 8 by 10%?
 (b) decreasing 8 by 10%?
 (c) increasing 15 by 100%?
 (d) increasing 24 by 20% and then decreasing the result by 20%?

1.13 Write each of the following numbers as a number in the range 1 to 10 multiplied by the appropriate power of 10:
 (a) 17.41, (b) 0.0317, (c) 15218, (d) 0.000147
 (e) 315.85×10^{-5}, (f) $3 \times 10^3 \times 2 \times 10^{-1}$, (g) $3 \times 10^3 + 2 \times 10^2$

1.14 Write each of the numbers listed in question (1.13) as a number in the range 0.1 to 100 multiplied by a power of 10 divisible by 3.

1.15 Write each of the following numbers as an ordinary decimal number (without a power of 10 as a factor):

(a) 6.732×10^4, (b) 2.155×10^{-2}

(c) 18.34×10^{-3}, (d) 0.4185×10^6

1.16 In what circumstances would it *not* be permissible to multiply both sides of an equation by $(x + 3)$?

1.17 *Without using a calculator*, determine which of the following integers are divisible by 2, 3, 5, 9 and 11:

(a) 26,   (b) 4345,   (c) 105

(d) 1179,   (e) 4649

1.18 What is the probability of obtaining no aces in a fair hand of bridge (i.e. 13 cards out of a total of 52)?

1.19 What is the probability of obtaining all cards of the same suit in a fair hand of bridge?

1.20 A class experiment to measure the degree of binding between two molecules showed a measurable effect when both concentrations were in the range 0.1–$5\,\mu\mathrm{mol\,L^{-1}}$. Which of the following values of the association constant calculated by several different groups of students could not possibly be correct?

(a) $3.73 \times 10^5\,\mathrm{M^{-1}}$,   (b) $8.5\,\mathrm{M^{-1}}$,   (c) $2.78 \times 10^{-3}\,\mathrm{M^{-1}}$,   (d) $5631\,\mathrm{M}$

1.21 The amino acid sequence of a protein is found to contain 150 residues, and the molecular mass under non-denaturing conditions is measured as 62 kDa. No carbohydrate, lipid or other co-factor could be detected. Is the native protein most likely to exist as a monomer, a dimer, a tetramer, or a hexamer?

CHAPTER 2

The language of Mathematics

2.1 **Introduction**

Lord Rutherford once remarked that all of science could be divided into physics and stamp collecting. This rather patronizing comment sums up what physicists often think about other sciences; biochemists are likely to find it unfair, but it has some truth in it. A science can hardly claim to be a science as long as it is just a catalogue of unrelated observations, and it is arguable that biochemistry has taken a step for the worse in the past couple of decades, when large parts of it have been taken over by an approach to molecular biology that at times seems little more than a vast catalogue of observations.

Only when general laws can be proposed and tested by experiment can a subject be said to have passed from mere description into science. In chemistry the transformation from stamp collecting into science corresponded with the development of thermodynamics and, later, the atomic theory and theories of chemical bonding; in biochemistry, the gradual realization that the understanding of life processes requires a foundation of physical chemistry and not just a list of metabolic reactions has played a corresponding role. It is no coincidence that mathematics has been central in all of these developments, and it is now almost impossible to grasp even elementary biochemistry without using elementary mathematics. Fortunately for non-mathematically minded biochemists, however, the mathematics necessary for an undergraduate course in biochemistry is not difficult, though it may appear so when met for the first time. Science students with any mathematical preparation in the years before university will have been exposed to most of it already, and those without such preparation can pick it up more easily than they often imagine. Little more is required than to identify the parts of elementary mathematics that are important in biochemistry and to reinforce them with appropriate examples.

Box 2.1 **From observations to equations**

A table tells us what was observed on one occasion, but it does not tell us what might have been observed under different conditions, e.g. what is the rate at a concentration of 8 mM?	Observations from a kinetic experiment	
	Substrate concentration (mM)	*Rate* ($\mu M\,s^{-1}$)
	1	2.00
	2	3.33
	5	5.56
	10	7.14
	20	8.33

An equation written in terms of numbers provides a summary of the table.

$$v = \frac{10a}{4+a}$$

It also allows predictions: $v = 6.67\,\mu M\,s^{-1}$ when $a = 8$ mM

An equation written in terms of symbols allows one to generalize from one set of observations to a law of how enzymes behave.

$$v = \frac{Va}{K_m + a}$$

In mathematics itself, the transformation from description into science is paralleled by the development from *arithmetic*, which is concerned with numbers and their manipulation, into *algebra*. Arithmetic is very useful, but it is not broad enough to satisfy all of the needs of science. In arithmetic, every problem is a new and separate problem, and it is difficult to make useful generalizations and hence to express scientific laws. Let us consider the simple biochemical example in **Box 2.1**, which shows a set of rates of a reaction measured at the substrate concentrations given. As it stands, the table is no more than a description of the results of a particular set of measurements and as long as we treat the numbers just as numbers we cannot make a general or useful statement about the enzyme to which they refer. The table tells us the rates observed at substrate concentrations of 5 and 10 mM but offers no guidance about what rate to expect at 8 mM; it does not tell us whether the system studied was behaving in accordance with some general law; it offers no clue as to what general law there might be. To remedy these omissions, we must move beyond arithmetic into algebra, because only then we will be able to recognize a pattern or regularity in the numbers, and express it so that it can be recognized again if it occurs with another system. If the rates in the table are represented as v and the substrate concentrations as a, then the equation below it expresses a law that defines all of the numbers in the table.

This equation has two advantages over the table: first, it allows a summary of all of the information while occupying much less space; secondly, it predicts what v values we might expect to observe at a values that are not included in the table. Suppose, for example, that we should like to know what the rate would be if a were 8 mM: by itself, the table tells us nothing about this, but the equation predicts that $v = 6.67\,\mu\text{M s}^{-1}$ when $a = 8\,\text{mM}$. The equation is thus more useful than just a list of numbers recorded on a particular occasion.

One can proceed one stage further with this example by noting that the equation is typical of what is reported for many enzymes, and so if we replace the numbers $10\,(\mu\text{M s}^{-1})$ and $4\,(\text{mM})$ by V and K_m respectively, we have an equation that expresses a generalization about enzymes, as shown in the equation at the bottom of **Box 2.1**. Again, replacing numbers with symbols has increased the generality of what we want to say. In addition to the symbols v, V, a and K_m, which represent numbers, either particular ones or generalized ones, the equation contains three operators: one is represented by the addition sign $+$; one is shown by the horizontal line between Va and $K_m + a$; and the third is implied by the juxtaposition of V and a, but could have been made explicit by writing $V \cdot a$ instead of Va. Each operator specifies something to be done to the numbers or symbols operated on: the $+$ sign requires K_m and a to be added together; the horizontal line requires Va to be divided by $K_m + a$; the juxtaposition of V and a (or a dot between them) requires them to be multiplied together.

This example ought to be familiar to you as the *Michaelis–Menten equation*, and its parameters V and K_m are called the *limiting rate* and *Michaelis constant* respectively. You may have seen V written as V_{max} and called the 'maximum velocity'; although this name is very common, it is a bad name from the mathematical point of view because the quantity that it defines is not a maximum in the mathematical sense, as we shall see in Chapter 4.

As long as no ambiguity is possible, mere juxtaposition is sufficient to indicate multiplication, but multiplication can be indicated by a dot or a cross if several numbers are to be multiplied together, or if we allow algebraic symbols consisting of more than one letter each (as in modern computer-programming languages), or if ambiguity is possible for some other reason. The dot is more common for pairs of symbols (where no confusion with the decimal point is possible) and the cross is more common for pairs of numbers, but the two symbols have the same meaning in most contexts, i.e. $V \times a$ means the same as $V \cdot a$. (In some specialized applications, such as in *vector algebra*, it is convenient to assign distinct meanings to the dot and the cross, but these need not concern us in elementary biochemistry.) In current usage the dot should be written *above* the line and the decimal point *on* the line, e.g. $5.1 \cdot 8.7 = 44.37$ *not* $5 \cdot 1 \cdot 8 \cdot 7 = 44.37$. This is a fairly recent convention so far as British books are concerned: although it has been used in American books for many years (certainly in all textbooks you are likely to come across), British books used exactly the opposite convention until comparatively recently (see **Box 2.3**).

Box 2.2 **Symbolism in computer languages**

Some older computer-programming languages did not require a multiplication symbol, but this was only possible because they insisted on very short symbols for variables	`A = AA`
Later on longer names appeared, but a compulsory multiplication symbol appeared at the same time	`RATESQ = RATE * RATE`
Good modern languages allow so much freedom in naming variables that separate explanations of what they mean are hardly necessary, and they also avoid using the = sign to mean 'replace'	`squaredRate:=Rate*Rate;`

Box 2.3 **Ways of writing numbers to be multiplied together**

A. *This book, American books of any age, British books since about 1980*

Decimal point on the line; multiplication sign raised; thin space to separate groups of three digits	$5.1 \cdot 8.7 = 44.37$ $1\,012.3 \cdot 13.1 = 13\,261.13$
The multiplication sign can always be written as a cross. This is uncommon in printed material, but less likely to be ambiguous in handwriting	$5.1 \times 8.7 = 44.37$

B. *Older British books (before about 1980)*

Decimal point *above* the line; multiplication sign *on* the line (exactly the opposite from above); comma to separate groups of three digits This convention is obsolete: *do not use it.*	$5\cdot1 \,.\, 8.7 = 44\cdot37$ $1,012\cdot3 \,.\, 13\cdot1 = 13,261\cdot13$

C. *Books published in France or Germany (even if they are in English)*

These often use a decimal comma, and may separate groups of three digits with a point on the line	$1.012,3 \cdot 13,1 = 13.261,13$

2.2 **Priority rules for operators**

To avoid ambiguity, it is important to realize that operators have to be obeyed in a proper order. Unlike ordinary language, equations are not read from left to right but in accordance with *priority rules* that require certain operators to be obeyed before others. Thus, the value of $5 \times 3 + 2 \times 4 - 3$ is 20, not 25 or 65, because multiplication must be done before addition or subtraction. In general, the rules are as follows: evaluate expressions within brackets first; if brackets are 'nested' (brackets within brackets), then evaluate 'inner' brackets before 'outer'; take powers before multiplying or dividing (work down from the top if there are powers within powers); and multiply and divide before adding or subtracting (**Box 2.4**).

It is always permissible and often desirable to use brackets to clarify an expression that might otherwise be misinterpreted. This is true even if the expression without brackets is strictly unambiguous. When in doubt (or if you think the reader may be in doubt) *use brackets*. It is perhaps worth noting that in mathematical usage the term 'bracket' can refer to any sort of bracket, whether (round), [square] or {curly}, and except when given special definitions they are all equivalent: $(x(2+y))$ means exactly the same as $[x\{2+y\}]$, etc. It is usual to choose round brackets for the inner-most set, with square brackets around them, and curly brackets around those: $\{[(\ldots)]\}$; however, this is not compulsory, and does not apply if any of the three has a special meaning. For example, in chemistry it is common to use square brackets for concentrations, e.g. [ATP] as the concentration of

---

Box 2.4 Priority rules

Evaluate expressions within brackets first	$2 \times (5 + 3) = 2 \times 8 = 16$	*NOT* $10 + 3 = 13$
Evaluate 'inner' brackets before 'outer'	$5 \times [6 + 3 \times (3+4)]$ $= 5 \times [6 + 3 \times 7]$ $= 5 \times [6 + 21]$ $= 5 \times 27 = 135$	*NOT* $5 \times 9 \times 7 = 315$
Take powers before multiplying or dividing	$5 \times 3^2 = 5 \times 9 = 45$	*NOT* $8^2 = 64$
With powers within powers, work down from the top	e^{-2x^2} means $e^{(-2x^2)}$	*NOT* $(e^{-2x})^2$
Multiply and divide before adding and subtracting	$3 \times 4 + 8/2 = 12 + 4 = 16$	*NOT* $3 \times 12/2 = 18$
Remember... There is *no priority rule* for multiple divisions in one expression	$48/12/2$? $= 48/(12/2) = 8$? or $(48/12)/2 = 2$?	This cannot be interpreted!

ATP, and when this is done it is done consistently, avoiding the simultaneous use of square brackets for simple algebraic purposes. In typography the word 'bracket' specifically means square bracket, a round bracket being a 'parenthesis' and a curly bracket a 'brace', but these terms are not usual in elementary mathematics: they are all just brackets, unless qualified as round, square or curly.

Example 2.1 Application of priority rules

What is the value of $y = 7 + 4/2 + (7 - 2) \cdot 6 - 5^2$?

The first step is to evaluate any expressions contained within brackets. $(7 - 2) = 5$, and so

$$y = 7 + 4/2 + 5 \cdot 6 - 5^2$$

Next any power must be evaluated. $5^2 = 25$, and so

$$y = 7 + 4/2 + 5 \cdot 6 - 25$$

Next come division $(4/2 = 2)$ and multiplication $(5 \cdot 6 = 30)$:

$$y = 7 + 2 + 30 - 25$$

Addition and subtraction are done at the end, when all the steps of higher priority have been completed:

$$y = 14$$

There are no rules of priority between addition and subtraction among themselves, because the result of a sequence of additions and subtractions is independent of the order in which they are done. This is normally a matter of convenience only, although sometimes numerical considerations may make one order better than another. In principle, the same applies to multiplication and division, but greater care is needed because thoughtless use of the slash / to indicate division often results in expressions with meanings that are either unclear or, worse, clear but different from what the writer intended. It is wisest therefore to use the slash in moderation and to check carefully that expressions have the meanings intended. Consider, for example, the following equation:

$$v = \frac{Va}{K_m(1 + i/K_i) + a}$$

This is unambiguous, and the priority rule should prevent the bracketed expression $(1 + i/K_i)$ from being misread as $[(1 + i)/K_i]$. When there is more than a single term after the slash, however, as in $(i/K_i + 1)$, misunderstanding is more likely because it is not always clear whether the slash indicates simple division according to the priory rule or whether it is used to avoid the typographical inconvenience of a cumbersome fraction such as $\frac{i}{K_i + 1}$, i.e. that the slash just replaces the horizontal line of the fraction.

Double slashes can be so confusing that they should *never be used*: this applies not only to ordinary algebraic expressions but also to the units of physical quantities, as in $R=8.314\,J/mol/K$. Here it is not clear whether the K belongs in the numerator of the unit with the J, or in the denominator with the mol, i.e. whether it means $(J/mol)/K$ or $J/(mol/K)$. To avoid this uncertainty the definition should be written as follows: $R=8.314\,J\,mol^{-1}\,K^{-1}$. For reasons that will become clear in Chapter 3, mol^{-1} and K^{-1} have the meanings $(1/mol)$ and $(1/K)$, respectively. In general, slashes should only be used in expressing units when there is only a single term in the denominator. People who use expressions like $J/mol/K$ will sometimes claim that there is no ambiguity because everyone knows what is meant and anyway it is obvious from chemistry that interpreting it as $J/(mol/K)$ would make no sense. True enough, as long as one is writing for readers with a thorough knowledge of chemistry; nevertheless, people who use such expressions sometimes do so also in cases where left-to-right evaluation does not give the right answer.

Computer-programming languages do not always obey precisely the same priority rules as conventional mathematics, with the result that now there is rather

---

Box 2.5 **Repeated operations**

Does it matter what order one does the same operation several times?

For adding and subtracting*, left-to-right is just the same as right-to-left (or any arbitrary order between these extremes):

	Left-to-right	Right-to-left
$6+3-5+4$	$[(6+3)-5]+4$	$6+[3+(-5+4)]$
	$=(9-5)+4$	$=6+(3-1)$
	$=4+4=8$	$=6+2=8$

The same is true for multiplying (without other operations mixed in)

$3\cdot2\cdot5\cdot6$	$[(3\cdot2)\cdot5]\cdot6$	$3\cdot[2\cdot(5\cdot6)]$
	$=(6\cdot5)\cdot6$	$=3\cdot(2\cdot30)$
	$=30\cdot6=180$	$=3\cdot60=180$

But it is not *true for division*

$12/2/3$	$(12/2)/3$	$12/(2/3)$
	$=6/32$	$=12/0.667=18$

Expressions with more than one division signs not separated by brackets are ambiguous, and *should never be written*.

*Strictly speaking subtraction is safe only if you regard every $-n$ (where n is any number or expression) as meaning $+(-n)$. You should never be tempted to interpret $6+3-5+4$ as $6+[3-(5+4)]=6+(3-9)=6-6=0$.

more confusion about the conventions than existed before computers became widespread. The appearance of cheap electronic calculators initially made matters much worse in this regard because those that use so-called 'algebraic notation' commonly ignored mathematical conventions altogether, using a 'left-to-right' system. (Calculators have, however, improved greatly in this respect since the first edition of this book was written, and one may hope that if there are future editions they may not need to mention this problem at all.) An expression such as

$$A * A / B / C / D + A / B * C$$

would be unambiguous in a computer program. (The multiplication sign is expressed here as $*$.) Most languages (Basic, Pascal, C, etc.) would give the expression the meaning

$$\frac{A \cdot A}{B \cdot C \cdot D} + \frac{A \cdot C}{B}$$

This unambiguous meaning, which may nonetheless be different from what the programmer intended, does *not* imply that such expressions are acceptable in ordinary mathematics. Similarly, the fact that simple calculators sometimes disregard priority rules does not mean that they are obsolete. For example, the expression $5 \times 3 + 2 \times 4 - 3$ must be interpreted as $(5 \times 3) + (2 \times 4) - 3$ if the proper conventions are followed, even though simple calculators using the so-called 'algebraic notation' may execute instructions as they are entered and consequently interpret the above expression as $\{[(5 \times 3) + 2] \times 4\} - 3 = 65$ (this is indeed what I get when I key in $5 \times 3 + 2 \times 4 - 3 =$ on a pocket calculator that came as a free gift with something I bought a few years ago).

2.3 **The summation sign**

It often happens, especially in statistical calculations, that we need to add together a large number of terms of the same kind. For example, if we have a set of values $x_1, x_2, x_3, \ldots, x_n$, their arithmetic mean $\bar{x}$ is given by

$$\bar{x} = \frac{x_1 + x_2 + x_3 + \cdots + x_n}{n}$$

In more complex examples explicit representation of the summation becomes cumbersome and unnecessary, and it is more convenient to use a special operator called the *summation sign* Σ (a capital Greek sigma) instead:

$$\sum_{i=1}^{n} x_i \equiv x_1 + x_2 + x_3 + \cdots + x_n$$

The *limits* $i = 1$ and n written above and below the summation sign mean 'start adding at $i = 1$ and continue until $i = n$'. If the limits are obvious, as for example in statistical calculations where one often has to sum over all of a set of observations, they can be omitted. The use of the *identity sign* $\equiv$ in this expression instead of the more common equals sign $=$ emphasizes that this is a definition: a general truth that does not depend on the particular values of x_1, x_2, etc. A case could be made for writing the previous equation, the definition of $\bar{x}$, as an identity, but it is a weaker case, because this is not the only way we could define $\bar{x}$. In general, mathematicians are fairly sparing in the use of identity signs, and expressions are often written as equations even if they could be regarded as identities.

....

Example 2.2 Evaluation of a sum

Evaluate $\displaystyle\sum_{i=1}^{4} x_i^2 - ix_i$, assuming $x_1 = 3.2$, $x_2 = 4.6$, $x_3 = 5.1$, $x_4 = 7.7$.

Writing out the summation in full, we have

$(3.2^2 - 3.2) + (4.6^2 - 2 \cdot 4.6) + (5.1^2 - 3 \cdot 5.1) + (7.7^2 - 4 \cdot 7.7)$

$= (10.24 - 3.2) + (21.16 - 9.2) + (26.01 - 15.3) + (59.29 - 30.8)$

$= 7.04 + 11.96 + 10.71 + 28.49$

$= 58.2$

....

Although the summation sign is a considerable convenience when the underlying calculation is understood, it can sometimes obscure the meaning when it is not. Indeed, one of the main reasons why more advanced mathematics can appear much more difficult than it is in reality is that it often uses special notation to express results more compactly. For example, the whole of *matrix algebra* is a way of expressing very complicated relationships in an extremely compact way: very convenient when one is familiar with the symbolism but baffling when one is not. Whenever obscurity threatens for this sort of reason, it is often helpful to translate the compact expressions into a more long-winded form, and then their meanings are likely to become clearer.

A corresponding sign for *products* also exists, although it is much less often encountered than the summation sign. It is written as the Greek capital pi, Π and is used in an exactly analogous way, i.e.

$$\prod_{i=1}^{n} x_i \equiv x_1 x_2 x_3 \cdots x_n$$

2.4 **Functions**

A mathematical *function* can be regarded as a set of instructions to carry out a series of operations on a variable or set of variables. For example, if we define v in terms of a as

$$v = \frac{Va}{K_m + a}$$

where V and K_m are constants, then we are defining v as a function of a, by defining what operations have to be carried out on a to obtain v. We can also have functions of more than one variable. For example, v may be determined not solely by a single concentration a but may depend both on a and on another concentration i:

$$v = \frac{Va}{K_m(1 + i/K_i) + a}$$

and now we say that v is a function of both a and i.

Sometimes we may wish to symbolize the existence of a dependence of one variable on another without specifying what the dependence is. We then often use the symbol $f(\ )$ or something similar, e.g.

$$v = f(a, i)$$

which states that v depends on a and i but does not indicate whether the dependence follows the equation given above or some other equation.

There are a number of functions that are so often required in mathematics that they are given special symbols. For example, if y is always given by multiplying x by itself, we say that is the *square* of x and symbolize the relationship as

$$y = x^2$$

where the superscript 2 indicates that we need to multiply $x \cdot x$. Conversely, x in this example is the *square root* of y, which we may write as

$$x = \sqrt{y}$$

or, more commonly and for reasons that I shall discuss in Chapter 3 (Section 3.2), we may express the same relationship as

$$x = y^{1/2}$$

or, especially in printed work where it is difficult to show fractions clearly:

$$x = y^{0.5}$$

Other functions of great importance are the *logarithmic* and *exponential* functions, which I shall also consider in Chapter 3, the *derivative*, or result of differentiation (Chapter 4), and the *integral* (Chapter 5).

There are others, such as *trigonometric functions*, that are important in mathematics generally, but have little application in elementary biochemistry, and so I shall omit details about them. On the other hand, there are certain functions that have little importance in mathematics as a whole but which are useful to define for biochemical purposes. For example, various properties of proteins can be related to the hydrogen-ion concentration $[H^+]$ in terms of the following kind of expression:

$$y = \frac{\tilde{y}}{1 + \dfrac{[H^+]}{K_1} + \dfrac{K_2}{[H^+]}}$$

in which $\tilde{y}$, K_1, and K_2 are constants. This kind of function is called a *Michaelis function*, because it was first studied by Leonor Michaelis, better known nowadays in biochemistry because of his association with elementary enzyme kinetics.

2.5 **Constants, variables and parameters**

Sometimes a quantity, such as the number 2.0, has a unique value under all circumstances and is called a *constant*. Other numbers, such as the *gas constant* $R \approx 8.3 \, \mathrm{J \, mol^{-1} \, K^{-1}}$ are found by experiment to be constant also. Others, such as K_m in the Michaelis–Menten equation,

$$v = \frac{Va}{K_m + a}$$

may be constant for a particular enzyme and substrate under well-defined and constant conditions, although they may vary with temperature, pH, etc. These quantities can be treated mathematically as constants only as long as the physical conditions that determine them are constant. Although K_m is called the Michaelis constant, and many similar quantities have the word 'constant' in their names, such quantities are constant in a much more limited sense than numbers like 2.0 or physical constants like the gas constant.

We are often interested in quantities that change when the conditions change. For example, we may find that an equilibrium 'constant' K varies with the temperature T according to the van't Hoff equation:

$$K = \exp\left(\frac{\Delta S^0}{R} - \frac{\Delta H^0}{RT} \right)$$

where exp( ) is the exponential function (Chapter 3), and ΔS^0, ΔH^0, and R are constants. Thus, although K may be constant at constant T, it varies with T. So, if we are concerned with changes in temperature we must treat K as a *variable*.

It will be evident that the distinction between a constant and a variable is not absolute, because any variable becomes a constant if the conditions that determine its variation are kept constant. In some contexts, the distinction is so difficult or inconvenient to make that we introduce a third term, *parameter*, to denote a quantity that is treated as variable for some purposes and constant for others. This occurs especially when we use statistical methods to estimate an unknown constant: although we may believe it to represent a true physical constant, we must treat it as a variable during the process of estimation. Suppose, for example, that we have a measured quantity y that depends on a controlled variable x according to the following linear equation:

$$y = a + bx + \varepsilon$$

in which a and b are the physical constants whose values we require and the third term ε on the right-hand side represents *experimental error*. This error term prevents us from calculating a and b exactly from measurements of y at two values of x. Instead, we would measure y at *several* values of x and try to find values of a and b that predicted the observed y values as closely as possible (I shall discuss a similar problem in more detail in Section 7.3). During the analysis we are free to try any values of a and b that we like: in other words, we treat them as variables, even though we may believe that some physical reality exists in which they have constant, but unknown, values. To express this duality we use the word *parameter* for a quantity such as a or b that controls the dependence between our observable variables.

2.6 **Dimensional analysis**

Most of the quantities we measure in biochemistry are not simple numbers but numbers with *units*. For example, the concentration of glucose in blood is not 0.005 but $0.005\,\mathrm{mol\,dm}^{-3}$. Put differently, and very formally, we may say that the blood-glucose concentration has the *dimensions* of amount of substance divided by length^3. Actually, this combination of dimensions occurs so often in biochemistry that it is convenient to define a *unit* of *amount-of-substance concentration*, such that 1 molar or $1\,\mathrm{M} \equiv 1\,\mathrm{mol\,dm}^{-3}$; and because amount-of-substance concentration is the only kind of concentration the biochemist is usually interested in, we usually abbreviate this cumbersome term to *concentration*. Strictly, it is a derived quantity, but it is often convenient to treat it as if it were a primary dimension.

Consideration of units and dimensions is sometimes regarded as a pedantic nuisance, but this is a pity, because it is one of the most powerful tools that scientists have for detecting mistakes in algebra, not only their own but also other people's. This is because there are rules that govern the way in which dimensioned quantities can be combined and a high proportion of algebraic mistakes cause these rules to be violated.

The simplest dimensional rule is that one cannot equate quantities of different dimensions: it is meaningless to assert that a length of 3 cm is equal to a mass of 1 kg, for example. This may seem so obvious as to be hardly worth mentioning, but it is surprising how many fallacies in kinetics spring from an inability to appreciate that one cannot compare a first-order rate constant with a second-order rate constant, because their dimensions are different. An obvious extension of the first rule is that one cannot add or subtract quantities of different dimensions or say that they are greater or less than one another. On the other hand, we can multiply them together or divide one by the other: if an object has a mass of 1 g and a volume of 2 cm^3, it is quite acceptable to divide 1 g by 2 cm^3 to obtain a density of $0.5\,\mathrm{g\,cm}^{-3}$.

Example 2.3 The rate of diffusion of a substance

The rate of diffusion of a substance in a gradient can be expressed as

$$\frac{dm}{dt} = -DA\frac{dC}{dx}$$

where dm is the mass (in g) transferred in a small time interval dt (in s), D is a constant called the diffusion coefficient, A is the area (in cm^2) of the plane through which the substance is diffusing, and dC is the change in concentration (in $g\,L^{-1}$, which is the same as $g\,dm^{-3}$) that occurs over a small distance dx (in cm) along the axis of diffusion. In what units must D be measured if this equation is to be consistent?

Left-hand side   *Right-hand side*

$$\frac{\text{mass}}{\text{time}} \qquad D \times \text{length}^2 \times \frac{\text{mass}}{\text{length}^3} \times \frac{1}{\text{length}}$$

To make the equation consistent, therefore, D must have dimensions of $length^2 \, time^{-1}$, and it is accordingly commonly measured in $cm^2\,s^{-1}$.

Note, however, that to do the calculation in terms of units rather than dimensions we would need to take account of the use of two different units of length (dm and cm) in the same expression. A second complication would arise if we chose to measure the concentration in $mol\,L^{-1}$ instead of $g\,L^{-1}$.

For certain mathematical purposes, only *pure numbers*, i.e. numbers with no dimensions, are admissible. For example, the expression 2^i means that i 2's have to be multiplied together, and this has meaning only if i is a pure number: it makes no sense, for example, to talk about multiplying 2 by itself 3 cm times. In other words, dimensioned quantities must not appear as *exponents*; for similar reasons their *logarithms* cannot be taken.

Sometimes we may wish to define a quantity, such as pH, that appears to be the logarithm of a dimensioned quantity, namely the hydrogen-ion concentration in the case of pH. However, as I discuss in Chapter 3, we can only do this if we first remove the dimensions by dividing by a *standard value* that has the same dimensions.

The use of all this is that when we make a mistake in algebra we often introduce a dimensionally incorrect expression. So, if we arrive at a result that we suspect may be incorrect, the simplest check we can make is for dimensional consistency. Suppose, for example, we are trying to remember the *Henderson–Hasselbalch equation* (Section 4.12), and think that it may be

$$pH = pK_a + \log([\text{salt}]\,[\text{acid}])$$

Is this likely? We can check by noting that [salt] and [acid] are both concentrations and therefore their product is a concentration squared. As this is not a pure

Box 2.6 **Rules for handling dimensions**

Don't equate quantities with different dimensions	$3\,cm = 1\,kg$ *not allowed*
Don't add or subtract quantities with different units	$10\,mol = 3\,cm + 1\,kg$ *not allowed*
When multiplying or dividing quantities with different units, multiply or divide the units as well as the numbers	$(5\,\mu mol)/(10\,mL)$ $= 0.5\,\mu mol/mL$ $= 0.5\,mmol/L$
Don't use a quantity with units as an exponent	$2^{5\,mL}$ *not allowed*
Don't take the logarithm of a quantity with units	$pH = -\log\,[H^+]$ *strictly not allowed, but …*
Whenever you see an apparent violation of the rule for logarithms there is an implied standard state	$pH = -\log\,[H^+]$ strictly means $pH = -\log\,\{[H^+]/[H^+]_0\}$

number, it cannot have a logarithm and so the expression must be wrong. If we are reasonably sure we have the right ingredients but that we have combined them incorrectly, we can use knowledge of the dimensions to deduce that we need a ratio of concentrations, which would be a pure number, not a product. But do we want [salt]/[acid] or [acid]/[salt]? Here dimensional analysis cannot help us, but our knowledge of chemistry can: we should know that the pH decreases as a solution becomes more acid, and this should guide us to the correct form of the equation:

$$pH = pK_a + \log\frac{[salt]}{[acid]}$$

If we have a long derivation that is dimensionally correct at the beginning but ends at a dimensionally incorrect result, we know that it must contain a mistake. How can we find it? One way might be to repeat the derivation, but then we may well repeat whatever thought processes led to the original mistake, in which case we shall not find it. A better approach—better because it is simpler, quicker and requires new thoughts rather than a repetition of old ones—is to check the dimensions at various stages until the mistake is located.

Dimensional analysis is also useful for remembering the slopes and intercepts of graphs. Anything measured along the x-axis, such as the intercept of a line on the x-axis, must have the same dimensions as x; anything measured along the y-axis must have the same dimensions as y. Furthermore, the *slope* at any point must have

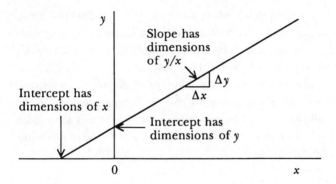

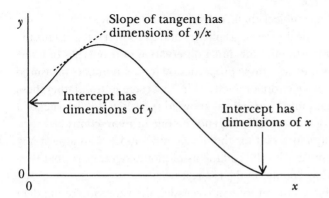

Fig. 2.1 Dimensional analysis as applied to graphs, for a straight line (*above*) and for an arbitrary curve (*below*).

the dimensions of y divided by those of x. The reason for this will become clear in Chapter 4 when we consider the definition of a slope as the limit of a change in the y coordinate divided by the corresponding change in the x coordinate as one moves along the line. For the present, we can accept it as a fact without worrying about the reason.

The rules for applying dimensional analysis to graphs are illustrated in Fig. 2.1. They allow us to check whether we have remembered the slopes and intercepts of particular plots correctly, and can guide us to the correct ones when we have not.

Some authorities advocate converting all variables into dimensionless form before plotting them. For example, instead of plotting [A], a concentration, in mM, we would make it dimensionless by dividing it by a 'standard' concentration of 1 mM, so that we actually plot [A]/mM, a dimensionless variable. Similarly, one can avoid the problems of defining thermodynamic quantities in terms of the logarithms of dimensioned equilibrium constants by defining all equilibrium constants in such a way as to make them dimensionless. My own view is that the advantages of following these recommendations are very slight and the disadvantages are considerable. If all variables are dimensionless, dimensional analysis becomes meaningless and one denies oneself the use of a valuable tool. One can also very easily be led into absurdities: by insisting that equilibrium constants are dimensionless, for example, one prevents the Michaelis constant K_m from being compared with the dissociation constant of the enzyme–substrate complex unless one defines K_m so that it is dimensionless as well; but doing this causes the Michaelis–Menten equation to contain a dimensionally incorrect addition of K_m to the substrate concentration—unless of course one is willing to have a dimensionless substrate concentration too! With all due respect to the distinguished scientists who have advocated this approach, it is hard to escape the feeling that it is a case of throwing out the baby with the bath water.

2.7 **Plotting graphs**

Biochemistry is a less visual subject than, say, zoology or botany; even by comparison with such mathematical subjects as physics and astronomy, it presents the experimenter with few opportunities for direct observation of properties of interest. Nearly always, we have to 'see' these properties by way of numbers measured on an instrument such as a spectrophotometer. The first stage in translating these rather abstract observations into biochemical knowledge is often the drawing of a *graph* to show how the observed variable depends on one or more controlled variables. It is therefore vitally important for the biochemist—more than almost any other scientist—to be familiar with the rudiments of plotting graphs so that they display the information they contain in the clearest way.

Rather than spend a large amount of text discussing the ways of plotting data well or badly, I have chosen a number of examples to illustrate common faults in plotting. First let us examine Fig. 2.2(a), where some observations are plotted with very poorly chosen scales on both axes. Less than a fifth of the ordinate scale is actually used, with the result that most of the graph is uninformative white paper. Expanding the ordinate scale, as in Fig. 2.2(b), improves matters considerably, but the large gap in data between $x = 20$ and 100 causes most of the observations to be crowded together on the left, so that there is still too much empty space. In fact most of the information is in these low-x data points and the one at $x = 100$ was probably measured just to give an idea of the limit approached by y. There is no real need to show the region between $x = 20$ and 100. We can then choose a scale that will spread out the low-x points, but break it to include the point at $x = 100$, as in Fig. 2.2(c).

Whenever a scale is broken as in this last example, the labelling must make it obvious that the break exists, to avoid misleading the reader. Figure 2.3 shows an example where this fault is taken one stage further by connecting the points across the break with a meaningless line: this is grossly misleading.

The graph illustrated in Fig. 2.4 *may* be quite legitimate; it will depend on what the objectives were when it was drawn. However, graphs that resemble it are often intentionally misleading and therefore dishonest. To the casual eye, the plot suggests that changing x brings about a substantial change in y. Only by checking the

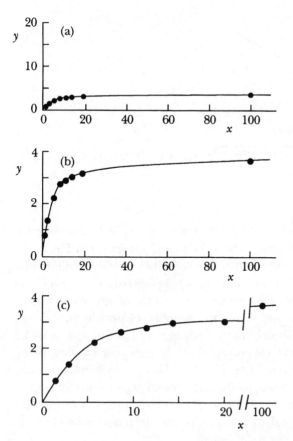

Fig. 2.2 (**a**) Poorly chosen scales result in compression of the data and hence loss of information. (**b**) Although choosing a better scale for *y* improves the plot, the large gap in the data between *x* values of 20 and 100 causes the majority of points to be squeezed against the ordinate axis. (**c**) Introduction of a clearly marked break in the scale for *x* allows the data points to be well spread out.

labelling does one see that the change in *y* is very small. Very much worse than what is shown here is to present the same sort of graph but to omit some or all of the labelling. Politicians and advertisers are fond of this kind of graph, but it has no place in science. Although one can sometimes find examples in the scientific literature, such graphs usually arise from carelessness or from failure to look at what is being done from the point of view of a reader rather than from deliberate dishonesty.

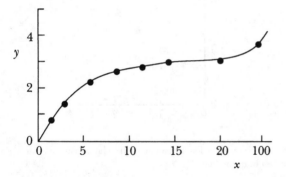

Fig. 2.3 Breaking the scale without marking it clearly is misleading, especially if the points are connected across the gap with a meaningless curve.

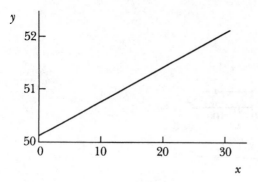

Fig. 2.4 Misleading axes can suggest a slope much larger than it really is.

The graph in Fig. 2.5(a) shows a very common fault: a straight line is drawn through points that demand a curve. The same points are shown in Fig. 2.5(b) without the line, and the curvature is somewhat more obvious: a line tends to bias the eye and it is prudent, therefore, to examine a set of points carefully *before* drawing a line through them. It is illuminating to take any issue of any research journal of biochemistry at random and search for examples of straight lines drawn through points that do not fit straight lines. They are not hard to find. A useful habit to cultivate is to examine all experimental graphs with your eye close to the paper and looking along the plotted line (Fig. 2.6). This greatly emphasizes any systematic failure of the points to lie on the line. In real experiments, of course, there is always some experimental error, so we should not expect an exact fit. But experimental error should be random, and so experimental points should be scattered randomly about the line, not systematically.

Looking at a graph as in Fig. 2.6 is a way of judging what the plot would look like if it were a *residual plot*, and it is a good idea to draw such a plot in reality

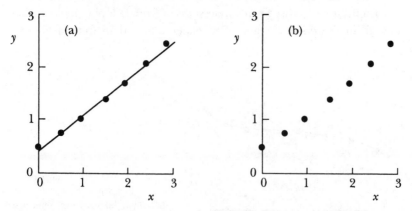

Fig. 2.5 (**a**) A straight line is drawn through points that demand a curve. (**b**) The curvature is more obvious when the same points are plotted without the line.

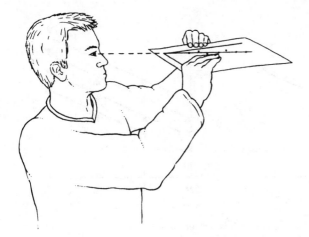

Fig. 2.6 Inspecting a graph with the eye close to the plane of the paper is helpful for noticing systematic deviations from the line.

whenever one has drawn a line that purports to fit a set of data. This is a plot of the differences between the points and the line against any convenient variable—often, though not necessarily, the same variable used for the abscissa in the original plot. Figure 2.7 shows a comparison between a normal plot, where the

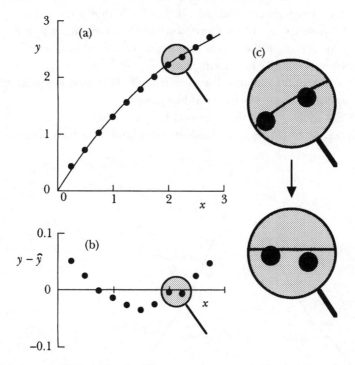

Fig. 2.7 (**a**) A normal plot makes it different to detect systematic deviations from the line. (**b**) However, these are magnified by plotting *residuals*, i.e. differences between the points and the fitted curve instead. (**c**) The magnifying glass shows what happens when (**a**) is transformed to (**b**).

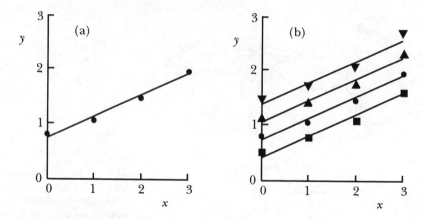

Fig. 2.8 (a) When only a few points are plotted it is difficult to judge whether an apparently systematic trend is due to chance. (**b**) However, if several lines show the same trend it is less easy to attribute it to chance.

systematic failure of the line to fit the points is not very obvious, and the corresponding residual plot, where the same failure can hardly be missed.

If there are only few experimental points, as in Fig. 2.8(a), it is very difficult to make a reasonable judgement about whether the scatter is random or systematic. The ideal answer is to obtain more observations, but this may not always be possible. Nonetheless, if the experiment is one of several in a series and they all show the same qualitative arrangement of points around the lines, as in Fig. 2.8(b), there is good reason to interpret it as systematic. It is usually legitimate in such cases to superimpose the separate residual plots.

Figure 2.9 shows a fault that seems to be particularly common in plots of pH dependences. Although any reasonable model would place the points on a smooth

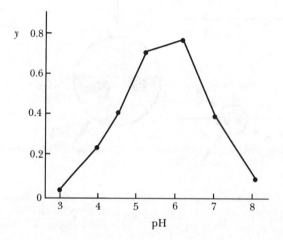

Fig. 2.9 If the points logically lie on a curve it is not usually helpful to connect them with straight lines.

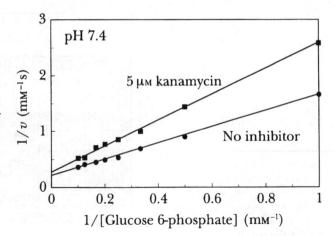

Fig. 2.10 An example of how to present data clearly. Notice that all of the information likely to be needed by the reader is present in the graph itself, making it unnecessary to refer to a separate list of symbols. This plot should be compared with Figs 2.11–2.13.

curve, they have been connected by straight-line segments. This is not only unattractive, it also misrepresents the form of the dependence, especially in regions of high curvature: in the Figure the maximum ought to be at pH 5.85, not at pH 6.1, and it should be at a height of 0.82, not 0.79. In experiments where there has been no attempt to account for the observations mathematically, it may sometimes be justifiable to connect points by straight-line segments simply to display them more prominently, but this should never be done in work with any mathematical pretensions.

Modern commercial graph-drawing packages have made it much easier to produce high-quality graphs, but they have also made it only too easy to produce

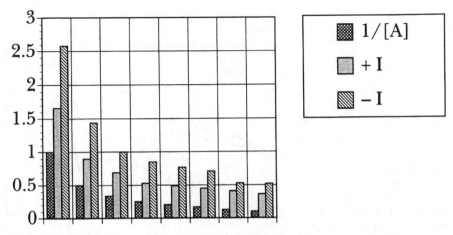

Fig. 2.11 Replotting the data of Fig. 2.10 in the form of a histogram makes the information virtually impossible to decipher.

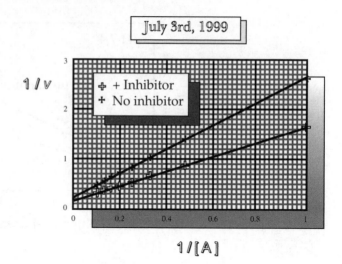

Fig. 2.12 Thoughtless use of the facilities provided by a commercial plotting package. The objective is to present the data clearly, not to show how many different styles of shading etc. are available.

perfectly horrible graphs, exhibiting faults that no one would have thought of 20 years ago. Figures 2.10–2.13 illustrate some of the possibilities, all based on the same sort of biochemical example. Figure 2.10 shows the kind of result that one can easily produce if one takes reasonable care: it is clear, readable and attractive, and includes all of the information needed for understanding the graph. Notice that the separate lines are labelled on the graph itself: this is much simpler for the reader than indicating it separately in the legend (e.g. ●, no inhibitor; ■, 5 μM kanamycin), though in graphs containing many lines one may be forced to consign this information to the legend to avoid excessive clutter. In the past, when graphs often had to be re-labelled by hand during publishing, journals sometimes

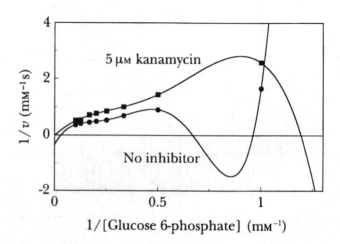

Fig. 2.13 Using a commercial line-fitting program allows one to fit all the points exactly, but the resulting curves are usually meaningless.

insisted on having lines identified in the legend rather than on the graph itself, partly to avoid errors, and partly to decrease the amount of work needed. However, in the modern context there is no need to sacrifice clarity for the sake of such minimal advantages.

Figure 2.11 illustrates how graphics packages offer many ways of plotting the same information, but these need to be used with discretion: the fact that you *can* plot kinetic data as a histogram does not make it a good idea to do so. Although the plot contains all the information (apart from units, which were omitted because the three variables plotted do not have the same units), it is almost impossible to decipher, because it fails to separate the two types of values (concentrations on the one hand and rates on the other).

Figure 2.12 illustrates (in an exaggerated form) what happens when one forgets that the object of a plot is to convey information, not to demonstrate how many different typestyles, plotting symbols and kinds of shading are available. Shadowing of any kind in a scientific graph is nearly always just meaningless clutter, and can make the scales difficult to read. The whole graph is dominated by a caption that almost certainly has no scientific importance. The labelling is likewise less helpful than it could be: rather than an anonymous 'inhibitor', why not specify the name of the inhibitor and its concentration? Neither of these improvements would normally add greatly to the amount of space taken up by the label. Most of what one can say here applies with even greater force to graphs prepared in colour. As a way of presenting information clearly, colour can be of great value, but colour for its own sake often just produces confusion.

In Fig. 2.13 we see what can happen when the powerful curve-fitting tools supplied in plotting packages are used without any understanding of their underlying assumptions. In a sense it shows the opposite fault from that in Fig. 2.5(a): there the problem was one of drawing a line that was too simple for the data, i.e. fitting the points too badly to be acceptable; here the problem is one of drawing lines that fit absurdly well (only too easy to do in the computer), ignoring any possibility of experimental error in the points, but following each one as if it were 100% accurate. Although in this case it is easy to see that the desire to fit every point has gone too far, it is unfortunately not always so easy to judge when enough is enough.

Box 2.7 **Using graphs to best advantage**

Don't compress scales unnecessarily

Avoid large empty regions

Don't change scales without making the change obvious

Make it obvious if the zero on any scale is not within the plotting region

Don't pretend the points fit a simpler line than they do

Don't connect points with meaningless lines

2.8 Precision

One of the commonest faults in scientific writing is to report results with an inappropriate degree of precision. For example, if a sample of 100 mL of blood is found to contain 93 mg of sugar, the concentration in mM (expressed in terms of glucose, which has molecular mass 180 Da) is $\frac{1000 \times 93}{100 \times 180}$, can be variously written as 5 mM (to one *significant figure*), or 5.17 mM (to three significant figures), or even 5.166 666 7 mM (to eight significant figures), etc. Which of these is appropriate for reporting the result? Commonsense tells us that 5 mM is rather more vague than the data would permit, whereas 5.166 666 7 mM is ridiculously precise, and 5.17 mM seems a reasonable compromise; but how can we justify such a compromise logically?

If we examine the four numbers used in the calculation, we see that 1000, the factor for converting from M to mM, is exact and we could write it as 1000.000 ... if we wished, 93 mg and 100 mL are of unknown accuracy, and 180 is correct to three significant figures, i.e. the true value is closer to 180 than to 179 or 181. Without knowing the conditions of the experiment, we cannot know how accurately the mass and volume were measured, but it seems reasonable to guess that the volume was chosen to be 100 mL by the experimenter and that it is correct to at least three significant figures. This leaves the 93 g as the probable main source of inaccuracy and the main one we have to consider: just as a chain is no stronger than its weakest link, the result of a calculation (in simple cases at least) is no more accurate than the most inaccurate value used in it. Presumably, the true mass was closer to 93 g than to 92 g or 94 g, but there is no reason to suppose it was 93.00 g rather than, for example, 92.68 g; but 92.68 g would give a calculated concentration of 5.148 888 9 mM (to eight significant figures) rather than 5.166 666 7 mM. Comparing these, we see that only the first two digits are reliable, though the third is not worthless: consequently 5.17 mM seems a reasonable way of expressing the concentration, with the understanding that the last (or *least significant*) digit is unreliable. To include any more than three significant figures in this result would be to claim that it is more accurate than it is. As a general rule, one should express a result with no more (or only a little more) precision than the accuracy of the data from which it was calculated.

Including too many significant figures may seem to be a harmless fault: after all, the reader can always ignore the unwanted digits. It is not harmless, however, because it can mislead the reader into thinking you have measured more than you have, and it can make you look a fool. Expressing the above concentration as 5.166 666 7 mM is just as silly as claiming that the fact that grass is green proves that it contains chlorophyll. The observation and the conclusion in this case are both true, and they are relevant to one another; nonetheless, the conclusion does not follow from the observation. Similarly, the true concentration of glucose could conceivably be 5.166 666 7 mM, but the available information does not justify saying so.

Inaccuracy can result not only from inaccuracy in the data but also from imprecise arithmetic during the calculation, especially if several steps are involved. It is advisable therefore to carry out the intermediate stages of a calculation with more precision (more significant figures) than one expects to retain at the end. However, it is not only very tedious to carry out every step with, for example, fifteen significant figures if only three are to be retained at the end; it is also unnecessary and it encourages mistakes. A fairly reliable rule is to use *one more significant figure* in the calculations than one expects to retain in the final result. This rule needs to be broken, however, in calculations that involve small differences between nearly equal numbers: for example, if a beaker weighs 10.381 4 g empty but 10.493 1 g after addition of a sample of a chemical it would be unwise to calculate the weight of chemical by subtracting 10.4 g from 10.5 g because this would give only one significant figure in the answer whereas the data would support four.

At the end of a calculation the unwanted digits have to be discarded. This can be done either by *truncation* or by *rounding*; thus 9.358 1 can be truncated to 9.35 or it can be rounded to 9.36. In truncation we simply omit the unwanted digits; in rounding we increase the last retained digit by 1 if the portion of the number to be discarded begins with 5, 6, 7, 8 or 9. Rounding is slightly more accurate than truncation, and is the usual practice in science.

In our own measurements we know (or should know) how precise they are, but in other people's there may be ambiguity. We had an example of this at the beginning of this section: the value of 180 Da given for the molecular mass of glucose is actually correct to three significant figures, but there is no way of knowing this from the way the number is written; it would still be written as 180 if a true value of 177 was for some reason being given to only two significant figures. If it is important to avoid this ambiguity we can state the precision explicitly, for example by giving the value as '180 (correct to three significant figures)', but this is cumbersome for ordinary use and we normally assume that we can gauge the precision by counting the number of significant figures. This is somewhat more subtle than it sounds, because although all non-zero digits are counted, zeroes may or may not be counted depending upon where in the number they occur. The simplest way of counting significant figures (which is easier to do than to describe) is to proceed as follows: (1) identify the first significant digit, which is the first non-zero digit encountered on reading the number from left to right; (2) identify the last significant digit, which is the last non-zero digit if the decimal point is not explicitly included, but the last digit of any kind if the decimal point is included; and (3) count all the digits (including zeroes), starting from the first and finishing with the last significant digit. This procedure may be illustrated by the following examples, in which significant zeroes are underlined and the number of significant figures is shown in parentheses after each number: 10_7 (3); 13.0_64 (5); 0.12_0 (3); 0.12 (2); 100 (1); 1_00.0 (4); 003 (1). (Zeroes are not usually written in front of integers, as in 003, which would normally be written simply as 3, but if they are written they are not significant.)

We do not usually mark significant zeroes in any special way (i.e. the under-lining done in these examples is not common practice) and it is cumbersome to add parentheses specifying the precision. There is, however, a third way of indi-cating that a value such as 100 mL is intended to have three significant figures: we can choose different units that make the precision clear, e.g. we can write it as 0.100 litre, or we can achieve the same result by relating the value to a power of 10 (see Section 3.1), e.g. by writing it as 1.00×10^2 mL. These representations are often used in scientific writing.

Statistical methods are discussed later in this book (Chapter 8), but mentioning them is appropriate here because they can provide estimates of the precision of calculated values with less of the guesswork and plausibility arguments that have characterized much of this section. When such a precision estimate exists, it can be written after the value prefixed by a $\pm$ sign ('plus-or-minus') thus: 5.17 ± 0.08 mM. There is rarely any point in reporting a precision estimate with more than one or two significant figures, and the value it qualifies should be reported with the same number of *decimal places*, i.e. the same number of significant digits after the decimal point. If there is no decimal point, the two numbers should have the same number of insignificant zeroes. Thus, we might write 81.031 ± 0.006 or $135\,200 \pm 1400$, but there would be little point in writing $135\,217 \pm 1400$ because the right-hand 17 would clearly be meaningless and so the number is better expressed as $135\,200$.

Finally, I should say a word about the difference between *accuracy*, which is con-cerned with the truth of what we say, and *precision*, which is concerned with how we say it. For example, the constant π has a value of about 3.141 59, and so the expression $\pi = 3.589\,37$ may be very precise but it is not at all accurate, whereas the expression $\pi = 3.14$ is much *less* precise but much *more* accurate. Precision is always within our control, but accuracy often is not. The main theme of this sec-tion has been that we should aim to express results so that their precision is only a little greater than their accuracy: less precision than this discards good inform-ation; more makes us look foolish.

2.9 Problems

2.1 If $x_1 = 1$, $x_2 = 3$, $x_3 = 2$, $x_4 = 7$, $x_5 = 6$, $x_6 = 9$, what are the values of the following expressions?

(a) $\displaystyle\sum_{i=1}^{6} x_i,$ (b) $\displaystyle\sum_{i=2}^{3} x_i,$ (c) $\displaystyle\sum_{i=1}^{4} i x_i,$

(d) $\displaystyle\sum_{i=1}^{6} x_i^2,$ (e) $\displaystyle\sum_{i=3}^{5} x_i / i,$ (f) $\displaystyle\sum_i x_i$

2.2 The Nernst equation asserts that

$$E = E^0 + \frac{RT}{n\mathrm{F}} \ln \frac{[\mathrm{ox}]}{[\mathrm{red}]}$$

where E and E^0 are potentials measured in volts, $R = 8.314\,\mathrm{J\,mol^{-1}\,K^{-1}}$ is the gas constant, T is the temperature in K, n is the number of electrons transferred in the oxidation–reduction process, F is a constant with the value $96\,494$ coulomb $\mathrm{mol^{-1}}$, and [ox] and [red] are the concentrations of the oxidized and reduced forms, respectively, of the redox couple. Assuming that $1\,\mathrm{volt} = 1\,\mathrm{J\,A^{-1}\,s^{-1}}$ and $1\,\mathrm{coulomb} = 1\,\mathrm{A\,s}$, show that the equation is dimensionally consistent.

2.3 Which of the following equations or statements must be incorrect because they are dimensionally inconsistent? In all cases, assume that [A], [I], K_m and K_i are concentrations; v and V are rates, i.e. concentrations divided by time.

(a) $v = \dfrac{V[\mathrm{A}]}{K_m + [\mathrm{A}] + [\mathrm{I}]/K_i}.$

(b) The intercept on the [A]/v axis of a plot of [A]/v against [A] is K_m/V.

(c) The slope of a plot of v against $v/[\mathrm{A}]$ is $-1/K_m$.

2.4 Evaluate the following expressions:

(a) $8 + 6/3 + 4$,   (b) $8 + 6/(3 + 4)$,   (c) $(8 + 6)/(3 + 4)$

(d) $2^3 + 3^2$,   (e) $2^3 \times 3^2$,   (f) 2^{3^2} (g) $(2^3)^2$

2.5 Different methods of determining the average molecular mass M_r of macromolecules in solution yield different kinds of average. For example, osmotic-pressure measurements yield a *number average* M_n, light-scattering measurements

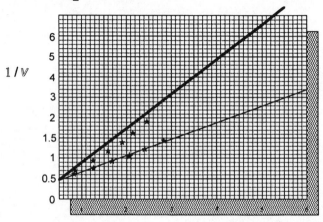

Fig. 2.14 See Problem 2.6.

yield a *weight average* M_w, and equilibrium sedimentation yields a *Z-average* M_z. These are defined as follows:

$$M_n = \frac{\sum n_i M_i}{\sum n_i}, \qquad M_w = \frac{\sum n_i M_i^2}{\sum n_i M_i}, \qquad M_z = \frac{\sum n_i M_i^3}{\sum n_i M_i^2}$$

where n_i is the number of mol of a species with $M_r = M_i$ and the summations are carried out over all solute species. Evaluate all three averages for a solution of protein existing as 27 nM monomer, 17 nM dimer and 1.3 nM tetramer, assuming that $M_r = 72$ kDa for the monomer.

2.6 The plot illustrated in Fig. 2.14 displays more than ten distinct faults, some of them important. List as many as possible.

2.7 Give the results of the following calculations with appropriate precision:

(a) $17.3 \times 1.382\ 21$,   (b) $2.571/(7.331 - 2.111)$
(c) $1.3359 \times (35.6579 - 35.6112)$,   (d) 2.1×4.361

2.8 How many significant figures does each of the following values contain?

(a) $23.007\ 050$,   (b) $0.007\ 050$,   (c) $135\ 000$
(d) 1.350×10^5,   (e) 10.37

CHAPTER 3

Exponents and logarithms

3.1 Integer powers

In mathematical manipulations one often has to multiply the same number by itself repeatedly, so that one has such expressions as

$$2 \times 2 \times 2 \times 2 \times 2 = 32$$

Such expressions rapidly become both tedious and difficult to read, and it is convenient to have a shorthand notation that conveys the same information not only more concisely but also (once one is used to it) more clearly. Exponents fulfil this need. The simplest use of an exponent i is to raise a number to its ith power, i.e. to show that the number is to be multiplied by itself i times, the exponent being written as a *superscript*, i.e. a little above the number that it operates on, and, usually, a little smaller in size. For example,

$$2^2 \text{ means } 2 \times 2 = 4$$
$$2^3 \text{ means } 2 \times 2 \times 2 = 8$$
$$2^4 \text{ means } 2 \times 2 \times 2 \times 2 = 16$$
$$3^3 \text{ means } 3 \times 3 \times 3 = 27$$

and so on. An exponent can have the value 1, but then it is usually omitted because the number it operates on is unaltered: $2^1 = 2$, $3^1 = 3$, $4^1 = 4$, etc.

It might seem that an exponent would have to be a positive integer, but if this were really so exponents would be much less useful than they are. In fact, it is possible to assign plausible meanings to zero, negative and fractional exponents and thereby to extend their usefulness very greatly. Consider the following pair of series:

$$i = 1 \quad 2 \quad 3 \quad 4 \quad 5 \quad 6 \quad \dots$$
$$2^i = 2 \quad 4 \quad 8 \quad 16 \quad 32 \quad 64 \quad \dots$$

What rules govern the choice of numbers to be put in each line? Each value of i is greater by 1 than the value on its left, whereas each value of 2^i is double the value on its left. What happens if we read the series from right to left? The rules can simply be reversed: each value of i is 1 less than the value on its right, and each value of 2^i is half the value on its right. There is nothing in these rules, however, that requires us to stop at $i=1$, and in fact we can continue applying them indefinitely (reading from right to left):

$$\ldots \quad -4 \quad -3 \quad -2 \quad -1 \quad 0 \quad 1 \quad 2 = i$$

$$\ldots \quad \frac{1}{16} \quad \frac{1}{8} \quad \frac{1}{4} \quad \frac{1}{2} \quad 1 \quad 2 \quad 4 = 2^i$$

I have used 2 as a base in these examples only because it is easy to do the calculations in the head, but the behaviour illustrated is not a special property of the number 2: we can get similar results for any number, not just 2, raised to the ith power, and in general we can define:

$$a^3 = aaa, \qquad a^2 = aa, \qquad a^1 = a, \qquad a^0 = 1$$

$$a^{-1} = \frac{1}{a}, \qquad a^{-2} = \frac{1}{a^2}, \qquad a^{-i} = \frac{1}{a^i}$$

The relationship $a^0 = 1$ must be noted particularly. It may seem surprising at first sight, but it arises quite naturally from the series that we have discussed.

The main use of integer powers greater than the second or third in biochemistry is for convenient representation of very large or very small numbers: we rarely need to write expressions like x^7, but we frequently need ones like 10^{-6} or 10^8. For example, the number of molecules in 1 mol of a chemical substance is about $600\,000\,000\,000\,000\,000\,000\,000\,000$, whereas the number of gram-ions of H^+ in a litre of 1 M NaOH is about $0.000\,000\,000\,000\,01$. Numbers such as these are very inconvenient to manipulate and even to read if they are written in the ordinary way. (This is perhaps borne out by the fact that in typing the manuscript of this book I needed three attempts at typing this last number before it was correct, and no doubt further care will be needed when the time comes to check the proofs.) In practice, therefore, it is usual to relate very large or very small numbers to the nearest power of 10: thus, the first of the two numbers mentioned is 6 times greater than 10^{23}, and is therefore usually written as 6.0×10^{23}, whereas the second is equal to 10^{-14}.

In practice, in experimental science we usually simplify even further, at least with dimensional quantities, by using multiples or submultiples of units: for example, the biochemist often has to deal with concentrations around 0.00001 M. This could be written as 10^{-5} M, but it leads to fewer mistakes in arithmetic if we use a smaller unit of concentration and write it as $10\,\mu$M, where the μ (read as 'micro' rather than 'mu' when it has this meaning) is a standard prefix meaning 10^{-6}. A table of the prefixes commonly needed in biochemistry is given as Table 3.1.

Table 3.1 Prefixes for powers of 10 with common biochemical applications.

Name	Symbol	Meaning	Example
mega-	M	$10^6 \times$	MV: megavolt
kilo-	k	$10^3 \times$	kg: kilogram
deci-	d	$10^{-1} \times$	dm: decimetre
centi-	c	$10^{-2} \times$	cm: centimetre
milli-	m	$10^{-3} \times$	mM: millimolar
micro-	μ	$10^{-6} \times$	μg: microgram
nano-	n	$10^{-9} \times$	nm: nanometre
pico-	p	$10^{-12} \times$	pg: picogram

The prefixes that refer to multiples of 10 that are not divisible by 3 (i.e. deci- and centi-) are normally used in biochemistry only for measurements of length, not for measurements of mass, volume, time, etc.

In principle, we could relate large and small numbers in this way to powers of any number, but it is normally convenient to use powers of 10 because 10 happens to be the basis of our ordinary number system. To avoid confusion, we always use powers of 10 even for numbers that have simple expressions as powers of other numbers. Thus, we would express 2401 as 2.401×10^3 even though it happens to be precisely 7^4.

There are three points worth noting in relation to the prefixes for powers of 10 listed in Table 3.1, as their neglect leads to confusion. The first point really refers to unit symbols in general and not just the prefixes, and is that they should never be printed in italics, to distinguish them from symbols for algebraic variables, which should nearly always be printed in italics. The second is that multiple prefixes should not be used: in the past, one often saw mμ (millimicron), itself an abbreviation for mμm (millimicrometre); this is far better written as nm (nanometre). The third point is that when a unit is raised to a power the prefix is considered to be part of the unit: thus $1\,\mu M^{-1}$ means $1\,(\mu M)^{-1}$, *never* $1\,\mu\,(M^{-1})$. Ignoring this last point can generate huge errors. Thus, $1\,\mu M^{-1}$ is $1/(0.000\,001\,M)$ or $1\,000\,000\,M^{-1}$, one million million times larger than the misinterpretation as $0.000\,001/(1\,M)$ or $0.000\,001\,M^{-1}$. If ever a calculation leads to a result that seems wrong by an absurdly large factor, it is worthwhile checking whether an error of this kind has been made. It is also worth remembering that even though $1\,\mu M$ is a much smaller unit than $1\,M$, $1\,\mu M^{-1}$ is a much *larger* unit than $1\,M^{-1}$.

Example 3.1 Converting units

The first example is almost trivially easy, but is of a kind required very frequently. The second is more typical of the sort of challenge one faces when reading up a new field, where the authors have used unfamiliar units and terminology, which need to be 'translated' into terms one can understand.

(a) If the reciprocal concentration of dithionite is $50\,M^{-1}$, what is its concentration, expressed in units such that the numerical value is between 0.1 and 100?

   Taking reciprocals of both number and unit, we get 0.02 for the number and M for the unit, i.e. $1/(50\,M^{-1}) = 0.02\,M$.

   However, 0.02 is not between 0.1 and 100, so we multiply it by 1000 and use a unit 1000 times smaller: $0.02\,M = 20\,mM$.

(b) The 'Pockels point' is a measure of the area occupied by one molecule in a monomolecular film. For saturated fatty acids, it is typically found to be $21\,Å^2$. The angstrom, a measure of length that is theoretically obsolete but still widely used by crystallographers and others, has a value $1\,Å = 0.1\,nm$. If 100 mg of a saturated fatty acid of molecular mass 0.4 kDa is spread on water as a monomolecular film, is this likely to cover a bath tub, a swimming pool, or a lake?

   The number of moles is the mass in g (0.1 g) divided by the molecular mass in Da (400 Da), i.e. $2.5 \times 10^{-4}\,mol$.

   Multiplying by the Avogadro constant $(6 \times 10^{23}\,mol^{-1})$ gives the number of molecules as 1.5×10^{20}. Each molecule occupies $21\,Å^2$, or $21 \times (0.1\,nm)^2$, i.e. $0.21\,nm^2$. So the total area is $1.5 \times 10^{20} \times 0.21\,nm^2$, or $3.15 \times 10^{19}\,nm^2$.

   This result is not easily visualized, however, and needs to be converted to more ordinary numbers. $1\,nm = 10^{-9}\,m$, so $1\,nm^2 = 10^{-18}\,m^2$ (remember that if the unit is squared the corresponding number has to be squared as well), so the area is about $31.5\,m^2$—a lot more than a bath tub, a lot less than a lake, but about right for a large swimming pool.

3.2 Fractional exponents

We have seen that the effect of increasing the exponent in an expression a^i by 1 is to cause the value of the expression to be multiplied by a; we shall now consider the effect of doubling the exponent. In this case we produce the *square* of a^i, i.e. $a^{2i} = (a^i)^2$. For example: $2^2 = 4$, $2^4 = 16 = 4^2$; $4^3 = 64$, $4^6 = 4096 = 64^2$, etc.

We can gain an understanding of fractional exponents by considering the reverse of this process: halving is the reverse of doubling, and taking the square root is the reverse of squaring. It seems reasonable, therefore, to define $16^{1/2}$ as 4, $4096^{1/2}$ as 64, or more generally $a^{1/2} = \sqrt{a}$. Similarly, if $a^3 = b$, then $b^{1/3} = a$; if $a^4 = b$, then $b^{1/4} = a$. In this way the logic can be extended until *any* number can be used as an exponent: for example, $2^{0.21} = 2^{21/100} = (2^{21})^{1/100} = 2\,097\,152^{1/100}$, or 1.1567 approximately.

If a is a positive number (not necessarily an integer), then a^i is meaningful for any value of i, as we have seen. But if a is negative, a^i has a simple meaning only if i is a positive or negative integer, and in this case the meaning is just the same as for positive a. For example:

$$(-2)^1 = -2$$
$$(-2)^2 = (-2)(-2) = +4$$
$$(-2)^3 = (-2)(-2)(-2) = -8$$
$$(-2)^{-1} = 1/(-2) = -1/2, \text{ etc.}$$

[I shall not consider fractional powers of negative numbers, such as $(-2)^{1/2}$, because, although there are circumstances in which it is useful to define such powers, these circumstances occur very infrequently in biochemistry: for the purposes of this book, therefore, we can insist that i must be an integer for a^i to have a meaning if a is negative.]

In the above set of examples, notice that $(-2)^2 = +4 = 2^2$, and in general, for any a, $(-a)^2 = a^2$. This raises the possibility of ambiguity in the definition of fractional powers. If we define $4^{1/2}$ as 'the number that when multiplied by itself gives 4', the definition is ambiguous because it is satisfied not only by 2 but also by -2. In most circumstances it is not at all convenient for a mathematical expression to have more than one meaning, and with fractional powers the ambiguity is removed by convention: $b^{1/2}$ is defined as the *positive* number a that satisfies the relationship $a^2 = b$. If we want to refer to the negative solution to the equation $a^2 = b$ we must write $-b^{1/2}$. Occasionally, we want to retain the ambiguity (as in the formula for the roots of a quadratic equation: see Chapter 6), and in this case we use the symbol $\pm$ ('plus or minus') to indicate that either sign may be used.

3.3 **Addition and subtraction of exponents**

We have seen that a^i means $a \times a \times \cdots \times a$, with i instances of a multiplied together, and similarly a^j means the same but with j instances of a. If these two products are multiplied together, the result must be the same as if we had multiplied $(i+j)$ values together at the outset, i.e. a^{i+j}. In general:

$$a^{i+j} = a^i a^j$$

Although it is only obvious that this result is correct for integer values of i and j, it actually applies for *any* values of i and j, including both fractional and negative ones. If j is negative—suppose $j = -k$, where k is positive—then the expression above gives

$$a^{i+j} = a^{i-k} = a^i a^j = a^i a^{-k} = a^i / a^k$$

Thus, just as *addition* of exponents corresponds to *multiplication* of powers, so *subtraction* of exponents corresponds to *division* of powers.

It is important to note that in all of the expressions in this section I have used the same base a: there are no correspondingly simple expressions for powers of different numbers. For example, if $a \neq b$, $a^i b^j$ does not have a simple expression in terms of added or subtracted exponents (unless of course b is a simple power of a, e.g. if $b = a^2$ then $a^i b^j = a^{i+2j}$).

3.4 **Logarithms**

Addition and subtraction are much quicker and easier operations to carry out than multiplication and division. The observation that addition and subtraction of exponents corresponds to multiplication and division of powers suggests a way in which one might use the operations of addition and subtraction to obtain the results of multiplication and division. For example, if one had a list of powers of 2, one might multiply 4 by 8 by noting that $4 = 2^2$ and $8 = 2^3$, so $4 \times 8 = 2^{2+3} = 2^5 = 32$. This is of course a trivial example as it stands, because 4 can be multiplied by 8 in the head more rapidly than one can consult a list of powers of 2. Moreover, if the list were confined to integer powers it would be unlikely to include the numbers we wished to multiply together. Suppose, however, that the list were extensive and included non-integer exponents: then we could use the same approach to solve non-trivial problems. Consider, for example, the product 2.38×9.23. An extensive table of powers of 2 would reveal that $2.38 \approx 2^{1.25}$ and $9.23 \approx 2^{3.21}$; therefore, $2.38 \times 9.23 \approx 2^{1.25+3.21} = 2^{4.46} \approx 21.97$. In this case the direct multiplication is not trivial and would take much more time than the corresponding calculation in terms of powers of 2—provided, of course, that suitable tables were available. In practice, it would be most convenient to have two sets of tables, one showing what exponent i is needed for 2^i to have a selected value, the second showing the value of 2^i for a selected value of i. These are known as tables of *logarithms* and *antilogarithms*, respectively. Instead of describing 2.38 as the result of raising 2 to the power 1.25, we can express the same relationship by describing 1.25 as the logarithm of 2.38 to the base 2. These relationships are symbolized as follows:

$$1.25 = \log_2 2.38$$
$$2.38 = \text{antilog}_2 1.25 = 2^{1.25}$$

Notice that 'antilogarithm' is simply another word for power: it is used to emphasize that finding an antilogarithm or a power is simply the reverse of finding a logarithm.

3.5 **Common logarithms**

It is not necessary to use 2 as the base for a set of logarithms: any positive number would in principle be usable, and it might seem that all positive numbers would be equally convenient to use. In fact, however, the two bases 10 and e (see Section 3.7) have considerable advantages over all other choices and are by far the most commonly used, although 2 is occasionally used in studies of bacterial growth, for reasons that I shall discuss later in this chapter (Section 3.8). The practical advantage of 10 over, say, 2 is apparent from the following comparison of $\log_{10}$ and $\log_2$ values for several numbers:

$$\log_2 2.741 = 1.454; \qquad \log_{10} 2.741 = 0.438$$
$$\log_2 27.41 = 4.776; \qquad \log_{10} 27.41 = 1.438$$

$$\log_2 274.1 = 8.098; \qquad \log_{10} 274.1 = 2.438$$
$$\log_2 2741 = 11.420; \qquad \log_{10} 2741 = 3.438$$

Notice that each multiplication of a by 10 increases $\log_2 a$ by $\log_2 10 = 3.322$, a rather unmemorable number, but it increases $\log_{10} a$ by $\log_{10} 10 = 1$ exactly. This means that if we use logarithms to the base 10, or *common logarithms* (also sometimes known as Briggsian logarithms, after Henry Briggs, the English mathematician who first published tables of logarithms in 1617), we need only a limited set of tables of $\log_{10} a$ for a values from 1 to 10, because all other common logarithms can be obtained from these by a trivial calculation. For example, if we required $\log_{10} 274.1$, we would look up $\log_{10} 2.741 = 0.438$ in a set of tables, note that 274.1 is $100 = 10^2$ times 2.741 and add 2 to the logarithm of 2.741, so $\log_{10} 274.1 = 2.438$.

Because of the separate ways of finding the integer and non-integer parts of a logarithm, they are given different names. The integer part (before the decimal point) is called the *characteristic*; the non-integer part (after the decimal point) is called the *mantissa*. Tables of common logarithms provide only the mantissa, as the characteristic is easy to determine by inspection. Conversely, tables of antilogarithms provide values only for the antilogarithm of the mantissa, the nearest power of ten being found by inspection of the characteristic.

Negative logarithms are the logarithms of numbers between 0 and 1, *not* the logarithms of negative numbers, which do not exist (see Section 3.6). It is common but not universal practice to write them with a positive mantissa and a negative characteristic. For example, $\log_{10} 0.2741 = -0.562$, but if it is written in this way it is not immediately obvious that it is exactly 1 less than $\log_{10} 2.741 = 0.438$. Instead, therefore, it can be written as $-1 + 0.438$ or, more concisely, as $\bar{1}.438$ (read as 'bar 1 point 438'), where the minus sign is written *above* the characteristic to show that it applies only to the characteristic and not to the mantissa. Similarly, $\log_{10} 0.002\,741 = -3 + 0.438 = \bar{3}.438$, etc. The need for this notation has decreased with the widespread availability of computers and electronic calculators, which have removed much of the drudgery from arithmetic, and have in particular made multiplication and division into trivial exercises. Consequently, one now has little occasion to write quantities such as $\bar{3}.438$, and one should especially avoid them in preparing data for plotting graphs, where they cause many mistakes. Nonetheless, the student should be familiar with the meaning of the notation as it is still encountered occasionally.

We have seen that $a^0 = 1$, regardless of the value of a. It follows that $\log_a 1 = 0$, regardless of a. The logic of this is evident from the fact that 0 is the only number that can be added to another number without affecting its value, just as 1 is the only number that can be multiplied by another number without affecting its value.

As we know that adding two logarithms together gives the logarithm of the product of the two numbers, it is worthwhile enquiring what we would get if the two numbers were the same:

$$\log a + \log a = \log(a \times a) = \log(a^2)$$

This is a particular case of a general relationship: multiplying a logarithm by a constant (2 in this case) is equivalent to taking the logarithm of the number raised to the power of that constant:

$$i \log x = \log(x^i)$$

Although the derivation is simple and obvious only in the case where i is a positive integer, this relationship actually applies to any value of i, including fractions, negative values, and zero. We can generalize it a little further by taking account of the property of sums of logarithms:

$$\log(x^i y^j) = i \log x + j \log y$$

This section and the preceding one can be summarized in the following set of equations:

$$\log(ab) = \log a + \log b$$
$$ab = \text{antilog}(\log a + \log b)$$
$$\log(a/b) = \log a - \log b$$
$$a/b = \text{antilog}(\log a - \log b)$$
$$\log(1/a) = -\log a$$
$$\log(a^b) = b \log a$$
$$a^b = \text{antilog}(b \log a)$$

The base of the logarithms is omitted from the symbols in all of these equations, because they apply for all bases, so long as the same base is used consistently: it is improper to mix logarithms to different bases in the same equation. The usual practice in biochemistry is to use the symbol log to mean $\log_{10}$ and ln to mean $\log_e$ (Section 3.7). *This practice is not universal*, however: pure mathematicians and others sometimes use log to mean $\log_e$.

3.6 **Negative numbers have no logarithms**

Each division of a number by 10 causes its logarithm to the base 10 to decrease by exactly 1. However, no amount of dividing by 10 can turn a positive number into a negative number, and so the original number remains positive throughout, even though its logarithm can cross zero and become negative. It is clear, therefore, that negative logarithms refer to small positive numbers, not to negative numbers. Negative numbers cannot be expressed as powers of positive numbers and, consequently, have no logarithms. This does not mean, however, that it is impossible to evaluate products that involve negative numbers. For example, -1.274×3.859 can readily be evaluated by use of logarithms by first evaluating 1.274×3.859 and then changing the sign of the result.

3.7 **Natural logarithms**

The second widely used kind of logarithms are called *natural logarithms* or *Napierian logarithms* (after John Napier, the 16th century Scottish mathematician), which have as base what may at first sight seem a bizarre choice, the number e. This number is *irrational*, which means that it cannot be expressed as an exact ratio of two integers, and consequently cannot be written exactly as a decimal number. It can, however, be expressed to any required accuracy as a decimal, and for most purposes it is more than sufficiently accurate to express its value as 2.718 28. (The repetition of 1828 in the more exact representation as 2.718 281 828 suggests an infinite repetition, but it is never wise to generalize after a single repetition, and this one does not continue.) One reason for choosing e as a base for logarithms is that in the days of calculation by hand it was easy to calculate its non-integer powers, by means of the following series:

$$e^x = 1 + \frac{x}{1!} + \frac{x^2}{2!} + \frac{x^3}{3!} + \frac{x^4}{4!} + \cdots$$

where the three dots imply that theoretically we must go on adding terms for ever, and the exclamation point indicates a *factorial*, i.e.

$$n! = n(n-1)(n-2) \cdots 3 \times 2 \times 1$$

for example,

$$6! = 6 \times 5 \times 4 \times 3 \times 2 \times 1 = 720$$

Although the series for e^x is strictly an *infinite* series, with an infinite number of terms, it is never necessary to evaluate more than a finite number of them, because their magnitude always dwindles into insignificance after the first few, unless x is very large (**Box 3.1**). If x is less than 1 this decay into insignificance is very rapid and for small values of x it is sometimes useful to consider only the first two terms in the series:

$$e^x \approx 1 + x$$

This is accurate to within $\pm 5\%$ if x is between -0.28 and $+0.35$: for example, if $x = 0.1$ then the approximation $e^{0.1} \approx 1.1$ is only 0.5% smaller than the correct value, 1.1052.

The infinite series provides a convenient way of calculating e^x, i.e. antilog$_e x$. Although in principle calculating $\log_e x$ is more difficult, a table of e^x values can be read inversely as a table of $\log_e x$ values; that is to say, if we have a table of e^x values and we can search in it for the particular x for which we want to know $\log_e x$. corresponding formula exists for calculating logarithms to any other base, such as 10. At the outset, therefore, it was much easier to obtain logarithms and anti-logarithms with e as base than with 10. Nonetheless, once a set of tables with base e is available, it is in principle simple to create a corresponding set with base 10, because for any number x there is a simple relationship between the logarithms to the different bases, as follows:

$$\log_e x = \log_e 10 \times \log_{10} x = 2.303 \log_{10} x$$

Box 3.1 **Infinite series for e^x**

The infinite series for e^x is $1 + \frac{x}{1!} + \frac{x^2}{2!} + \frac{x^3}{3!} + \frac{x^4}{4!} + \cdots$ (see text). Let us see how this behaves for a particular value of x, such as 3:

i	$x/i!$	$x(i-1)! \times 3/i$	$x/i!$	Sum
0	1		1.0000	1.0000
1	$x/1!$	$1.0000 \times 3/1$	3.0000	4.0000
2	$x^2/2!$	$3.0000 \times 3/2$	4.5000	8.5000
3	$x^3/3!$	$4.5000 \times 3/3$	4.5000	13.0000
4	$x^4/4!$	$4.5000 \times 3/4$	3.3750	16.3750
5	$x^5/5!$	$3.3750 \times 3/5$	2.0250	18.4000
6	$x^6/6!$	$2.0250 \times 3/6$	1.0125	19.4125
7	$x^7/7!$	$1.0125 \times 3/7$	0.4339	19.8464
8	$x^8/8!$	$0.4339 \times 3/8$	0.1627	20.0091
9	$x^9/9!$	$0.1627 \times 3/9$	0.0542	20.0633
10	$x^{10}/10!$	$0.0542 \times 3/10$	0.0163	20.0796
11	$x^{11}/11!$	$0.0163 \times 3/11$	0.0044	20.0840
12	$x^{12}/12!$	$0.0044 \times 3/12$	0.0011	20.0851
13	$x^{13}/13!$	$0.0011 \times 3/13$	0.0003	20.0854
14	$x^{14}/14!$	$0.0003 \times 3/14$	0.0001	20.0855
15	$x^{15}/15!$	$0.0001 \times 3/15$	0.0000	20.0855

Notice how although the terms initially get bigger, they start to get smaller as soon as the number in the denominator is bigger than x, and once that happens they decrease in size very fast and become negligible after a few more terms.

Of course, with an integer like $x=3$ we can calculate e^3 more rapidly as $2.718\,28 \times 2.718\,28 \times 2.718\,28 = 20.0855$, with the same result, but the great advantage of the infinite series is that it applies to *any* value of x, not just integers.

Moreover, if x is quite small, the terms get smaller (or, as a mathematician would put it, 'the series converges') much faster than in the example. Try it with $x=0.1$: you should find that the result $e^{0.1} = 1.1052$ can be calculated (to four decimal places) with just four terms. Then, with a few small values of x we can calculate any others we like by applying the standard properties of logarithms. For example, if we know $e^{0.1} = 1.1052$ and $e^{0.2} = 1.2215$ it is easy to calculate $e^{0.3} = e^{0.1+0.2} = 1.1052 \times 1.2215 = 1.3500$ without using the infinite series at all.

As we have seen, however, 10 is much the most convenient base for purposes of calculation: why then are logarithms to base e of more than just curiosity interest?

The continued importance of e and its functions such as $\log_e x$ and e^x does not, in fact, derive from their almost non-existent use as aids to arithmetic but from the fact that they occur naturally in the solutions to many kinds of problems in applied mathematics. We shall encounter various examples of this in this book and at this point it will suffice to indicate two briefly:

(1) In a system at thermal equilibrium at an absolute temperature T, the numbers n_1 and n_2 of molecules in two states with energies E_1 and E_2 respectively are related according to the equation

$$\frac{n_1}{n_2} = e^{(E_2 - E_1)/kT}$$

in which k *is* a constant known as the Boltzmann constant $(= 1.38 \times 10^{-23}$ $J K^{-1} = R/N$, where $R = 8.31 J K^{-1} mol^{-1}$ is the gas constant and $N = 6.02 \times 10^{23} mol^{-1}$ is the Avogadro constant). If we take natural logarithms of both sides we can express the difference between the energies as

$$E_2 - E_1 = kT \log_e(n_1/n_2)$$

(2) In a first-order reaction with rate constant k, the extent of reaction after time t is proportional to $1 - e^{-kt}$.

Although I have deliberately chosen rather physical examples here to illustrate the ubiquity of e, the fundamental reasons for the ubiquity are mathematical: they derive from the natural occurrence of $\log_e x$ and e^x in purely mathematical contexts, such as calculus (Chapters 4 and 5).

Powers of e occur so often that it is typographically inconvenient to write them as e^x, especially if x represents a complicated expression. Accordingly, the special operator exp is commonly used instead, i.e.

$$\exp(x) \equiv e^x$$

the expression being read as 'exponential x': it is unnecessary to specify the base because e is universally understood in this context. Using this notation we could have written the equation in the first example above as

$$\frac{n_1}{n_2} = \exp\left(\frac{E_2 - E_1}{kT}\right)$$

thereby avoiding the use of very small type, which can be difficult to read. Indeed, if we wanted to make this as easy as possible to print without making it difficult to read we could write the two fractions on the line, as follows:

$$n_1/n_2 = \exp[(E_2 - E_1)/(kT)]$$

If we do this, we must take care to use brackets to show that the whole of $E_2 - E_1$ is to be divided by the whole of kT and that the exp symbol acts on the whole fraction. (Many authors would feel that the brackets around kT are sufficiently obvious to be *understood*, i.e. to be omitted, but it never does any harm to include unnecessary brackets.) In handwriting, it is best to write fractions as in the first example, as it is no more difficult than writing them on the line, and it both avoids excess brackets and makes it easier for the reader.

Although in this section I have used the symbol $\log_e x$ for the natural logarithm of x, the usual practice in biochemistry is to write it as $\ln x$ (as already noted) and this practice is followed elsewhere in the book.

As mentioned, in the past it was much easier to calculate natural logarithms than common logarithms, but the latter were much easier to calculate *with*, and as a result both kinds of logarithms continued to be used. Nowadays, almost everyone has ready access to a calculator that can instantly provide either kind of logarithm, and either kind of antilogarithm, at the touch of a key, and logarithms are now rarely used anyway as an aid to multiplication and division. We may expect, therefore, that common logarithms have outlived their usefulness and will gradually cease to be 'common'.

3.8 **Logarithms to base 2**

We have seen earlier in this chapter that logarithms to base 2 provide a convenient and easily understood introduction to the idea of logarithms and their arithmetic properties. Although for most practical purposes 10 used to be the easiest base to work with and e is usually appropriate in theoretical discussions, 2 has an important use as a base in bacteriology.

When bacteria grow in ideal conditions in an open system provided with ample supplies of all necessary nutrients, the rate of growth is determined only by the metabolic capability of each individual bacterium to grow and divide. In a homogeneous bacterial culture there are no qualitative differences between individuals and quantitative differences are slight. Consequently, it is meaningful to conceive of the *generation time*, i.e. the period between one cell division and the next, as essentially constant from one cell to another. If the 'initial' population size (i.e. the number of cells at an arbitrarily defined time zero) is n_0 and the generation time is τ then at

$$t = 0, \quad n = n_0$$
$$t = \tau, \quad n = 2n_0$$
$$t = 2\tau, \quad n = 4n_0$$
$$t = 3\tau, \quad n = 8n_0$$

and, in general, at

$$t = z\tau, \quad n = 2^z n_0$$

Taking logarithms (to any base) of the expression for n, we have

$$\log n = z \log 2 + \log n_0$$

which we can rearrange into an expression for z:

$$z = \frac{\log n - \log n_0}{\log 2} = \frac{\log(n/n_0)}{\log 2}$$

the second form following from the ordinary properties of logarithms that we have already considered. As we know that $t = z\tau$, and hence that $z = t/\tau$, we can treat this last equation as an expression for t/τ:

$$t/\tau = \frac{\log n - \log n_0}{\log 2} = \frac{\log (n/n_0)}{\log 2}$$

This result is independent of the base of logarithms (so long as it is consistent throughout), but it assumes a particularly simple form if we choose a base that makes $\log 2$ take a value of 1, as we can then omit $\log 2$ from the equation (dividing by 1 being a trivial operation that leaves the value unchanged). The only base in which $\log 2$ has a value of 1 is 2, i.e. $\log_2 2 = 1$, but $\log_{10} 2 \neq 1$, $\log_e 2 \equiv \ln 2 \neq 1$, etc. So we can write

$$t/\tau = \frac{\log_2 n - \log_2 n_0}{\log_2 2} = \frac{\log_2 n - \log_2 n_0}{1} = \log_2 (n/n_0)$$

The phase of growth in which the mean generation time is a constant is known properly as the *exponential growth phase*. Paradoxically, the term *logarithmic growth phase* is synonymous, even though it sounds as if it means the opposite. The reason for this peculiar terminology is that during exponential growth the population size is not a linear function of time but can be plotted as a straight line if its *logarithm* is plotted against time (Fig. 3.1).

Notice that as the generation time τ is the time it takes for the population size to double, it can be, and often is, called the *doubling time*.

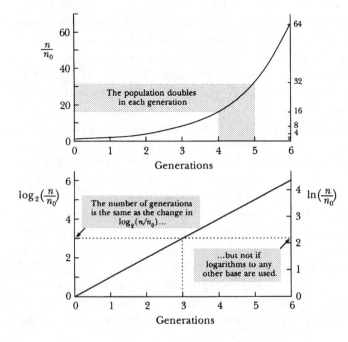

Fig. 3.1. Logarithms to base 2 provide a convenient way of relating population size to generation number, for populations that double in size in each generation.

3.9 **Exponential decay**

Although exponential growth is observed mainly with bacteria, exponential *decay* is of great importance to the biochemist in at least two other contexts: radioactive decay and chemical reactions that display first-order kinetics. Mathematically, all of these processes are very similar, differing only in the sign of the coefficient of time in the exponent, and the mean generation time in a growth process is paralleled by the *half-life* of a radioisotope or reactive chemical and the *half-time* of a chemical reaction. In principle, logarithms to base 2 might be used in these contexts but in practice they are not. The reason for this is partly historical: methods for analysing decay processes were already well established before 1949, when Monod drew attention to the usefulness of 2 as a base; but more important is the fact that the mean generation time of a bacterial population has a tangible physical meaning whereas the half-life of a radioisotope is only a mathematical abstraction, albeit a convenient one. Thus, there is a real interest in knowing the mean number of generations in a period of bacterial growth, whereas the number of times that a sample has lost half of its radioactivity is no more fundamental than the number of times its activity has decreased by a factor of e or by a factor of 10.

3.10 **Logarithms as a method of scaling**

In the chemistry of processes in aqueous solution the concentrations of hydrogen ions that need to be considered extend over more than 14 orders of magnitude, from less than 10^{-14} M in strong alkali to more than 1 M in strong acid. In the chemistry of living systems, this range is considerably compressed, but it is still large enough to make it arithmetically inconvenient to work exclusively in concentrations. In addition to the numerical inconvenience of concentrations, there is a philosophical reason for avoiding them. Biochemistry, like much of chemistry, tends to be more interested in changes in state than in states, and a particular change in concentration can have very different effects in different contexts. For example, in the mammalian stomach at a hydrogen-ion concentration of about 10^{-3} M an increase of 10^{-5} M would virtually be imperceptible and would be expected to have only slight consequences; on the other hand, an increase of 10^{-5} M in the hydrogen-ion concentration in a cell at an initial concentration of 10^{-7} M might well be devastating. (This is not a wholly abstract example: the increase in acidity brought about by rupturing lysosomes has major effects on the activities of numerous enzymes within the cell but would pass unnoticed in the stomach.) The point is that in the first case the change is 1.01-fold whereas in the second it is 100-fold: in most circumstances it is equal *relative* changes that bring about similar effects. It would be an oversimplification to say that a doubling (for example) of hydrogen-ion concentration has the 'same' effect when applied to starting values of 10^{-3} and 10^{-7} M, but such a statement makes sense conceptually and is not without meaning. On the other hand, it is difficult to think of circumstances in which it would make sense to regard increases of 10^{-5} M as the 'same', whether from 10^{-3} or from 10^{-7} M.

The purpose of this digression has been to show that numerical convenience is not the only reason for introducing logarithms into consideration of hydrogen-ion concentration. Not only does the conversion of $[H^+]$ into $\log_{10}[H^+]$ avoid the need for inconveniently small numbers; it also provides a scale in which particular linear changes (in $\log_{10}[H^+]$) have approximately constant meanings. Unfortunately, the scale that is in common use is not $\log_{10}[H^+]$ but $-\log_{10}[H^+]$, generally written as pH. For the ludicrously trivial 'advantage' of having a range of positive numbers around $+7$ rather than negative numbers around -7, chemists have paid the excessive price of discarding a scale of obvious meaning in favour of one that generates endless confusion. However, the pH scale is probably here to stay, and so one must make the best of an error of judgement and try to remember that pH is nothing more than a tiresome way of writing $-\log_{10}[H^+]$. Similarly K_a means $-\log_{10}K_a$, where K_a is an acid dissociation constant. Other usages, such as pK_m (where K_m is the Michaelis constant) are less common: I trust that they will remain so and that the time will never come when it is thought helpful to overcome the 'inconvenience' of having a minus sign in the definition of the standard Gibbs energy

$$\Delta G^0 = -RT \ln K$$

by writing it as $\Delta G^0 = 2.3RTpK$. We are fortunate that at least some extensions of the pH idea, such as the rH scale for application to redox potentials, have virtually disappeared.

3.11 Products of equilibrium constants

In a set of equilibria for a sequence of reactions, such as the following sequence of unimolecular steps:

$$A \underset{}{\overset{K_{AB}}{\rightleftharpoons}} B \underset{}{\overset{K_{BC}}{\rightleftharpoons}} C \underset{}{\overset{K_{CD}}{\rightleftharpoons}} D \underset{}{\overset{K_{DE}}{\rightleftharpoons}} E \underset{}{\overset{K_{EF}}{\rightleftharpoons}} F$$

the equilibrium constant for the complete process is the result of multiplying all of the individual equilibrium constants together:

$$K_{AF} = K_{AB}K_{BC}K_{CD}K_{DE}K_{EF}$$

Metabolic pathways are composed of such sequences of steps (although many are more complex than unimolecular steps) and it is frequently necessary to carry out such multiplications and corresponding divisions in order to assess the thermodynamic feasibility of particular pathways or importance of intermediates, etc. But it is much easier to think in terms of addition and subtraction than in terms of multiplication and division; for this reason, it is often useful to express equilibrium constants on a logarithmic scale, so that

$$\log K_{AF} = \log K_{AB} + \log K_{BC} + \log K_{CD} + \log K_{DE} + \log K_{EF}$$

Now it happens, for reasons that are primarily related to chemistry rather than to mathematics, that there is a direct relationship between the equilibrium constant K of a reaction and the energetic characteristics of the same reaction. Because of this relationship, it is common to use not the simple logarithm to base 10, which would be best if combining equilibrium constants were the only concern, but instead the natural logarithm multiplied by a temperature-dependent coefficient, i.e. the standard Gibbs energy ΔG^0, defined as follows:

$$\Delta G^0 = -RT \ln K = -2.3RT \log_{10} K$$

where $R = 8.31 \, \mathrm{J \, mol^{-1} \, K^{-1}}$ is the gas constant and T is the temperature in kelvin. Nonetheless, although ΔG^0 has a fundamental thermodynamic meaning, the factor $-2.3RT$ is irrelevant to many of the biochemical contexts in which a logarithmic scale is useful. Living systems hardly ever exploit temperature effects (they usually try to avoid them altogether by seeking out or creating conditions in which their cells are maintained at constant temperature); so, in metabolism $-2.3RT$ is just a constant, with a value of about $-5700 \, \mathrm{J \, mol^{-1}}$ (at 25 °C). Thus each $-5.7 \, \mathrm{kJ \, mol^{-1}}$ in a ΔG^0 value corresponds to a factor of 10 in the corresponding equilibrium constant.

3.12 Logarithms of dimensioned quantities?

It is obvious from the simplest ways we use exponents that they must be pure numbers, i.e. they must not have any dimensions: for example, it makes no sense to say that 10 is to be multiplied by itself 3 cm times. As a logarithm is the exponent that the base must be raised to in order to give a particular number, it follows that a logarithm can also have no dimensions. Moreover, a pure number raised to a pure-number power must also be a pure number, and so only pure numbers have logarithms, because taking a logarithm is the reverse of raising a pure number to a power. (For a fuller account of this argument, see **Box 3.2**.)

Despite this, there is apparently a glaring exception, one that has already occurred in this chapter: if we define pH as $-\log_{10}[H^+]$, and we measure $[H^+]$ in M then we are apparently taking the logarithm of a concentration. The way out of this difficulty is to define a *standard state*, in this instance a reference concentration $[H^+]^0$, and to define pH as $-\log_{10}([H^+]/[H^+]^0)$. In this way, pH becomes the logarithm of a dimensionless ratio, as it should be. In practice, the standard state is always taken to be $[H^+]^0 = 1 \, \mathrm{M}$, and this makes the ordinary definition of pH numerically correct *provided that* $[H^+]$ is measured in M (or $\mathrm{mol \, L^{-1}}$, which is the same: see **Box 1.5**).

This is a very important proviso to understand, because in principle we can measure $[H^+]$ in any units we find convenient—in tons per cubic foot if we so

---

Box 3.2 **Why must a number that has a logarithm be dimensionless?**

$[ATP]^3$ means $[ATP][ATP][ATP]$: three concentrations multiplied together, but what would $[ATP]^{3\,cm}$ mean?

It makes no sense to speak of multiplying 3 cm concentrations together, and so expressions like $[ATP]^{3\,cm}$ are considered to be meaningless.

In other words, and more generally, a dimensioned quantity cannot be an exponent.

If we write $\log_{10}(x) = y$ this implies (by definition) that $10^y = x$. However, we have just seen that in an expression like 10^y the exponent y must be dimensionless, and regardless of its value the value of 10^y will be a dimensionless number, hence x must also be dimensionless.

In other words, and more generally, a dimensioned quantity cannot have a logarithm.

---

wish—but the shorthand definition of pH with the standard state omitted gives the right answer *only* if the units of $[H^+]$ are M, which in practice means that it is only safe to use this kind of shorthand if there is no doubt about what units are implied. The fuller definition of pH may seem pedantic and trivial, but it is important for understanding that although pH differences have a fundamental meaning, pH itself is measured from an arbitrary datum and therefore has no absolute meaning. Thus, although it makes good sense to say that a pH change of $+0.2$ is twice as large as a change of $+0.1$, there is no useful sense in which pH 6 can be regarded as twice as alkaline (or even as twice as 'large', where the definition of large is kept suitably vague) as pH 3. Suppose, for example, that we redefined the standard state as $[H^+]^0 = 10^{-7}$ M (or 0.1 µM, a datum that would make better biochemical sense than the chemists' 1 M): pH 3 would become $pH' = -4$ and pH 6 would become $pH' = -1$, a quarter as 'large' instead of twice as 'large'! On the other hand, a change in pH of $+0.2$ would also be a change in pH' of $+0.2$ and would still be twice as large as a change of $+0.1$, because the definition of the standard state does not enter into measurements of change.

3.13 **Metabolic 'efficiency'**

Not many biochemists would suppose that pH 6 could meaningfully be regarded as twice as large as pH 3, but equally ridiculous comparisons are commonplace in discussions of ΔG^0 values for metabolic processes. Standard states do not enter into the definitions of dimensionless equilibrium constants (as in the reversible unimolecular reactions considered in the preceding section), but many metabolic and other reactions involve changes in the number of molecules and have equilibrium constants with dimensions that must be removed by the use of standard states. As in the definition of pH, the commonest convention is to define dimensionless equilibrium constants by relating concentrations to standard concentrations of 1 M. As a consequence of this introduction of standard states, it is usually meaningless to measure the 'efficiency' of a metabolic process as a ratio of ΔG^0 values.

Unfortunately, this is an error that is not only made by students but examples can also be found in quite respectable textbooks. We need, therefore, to look at it in a little more detail. The reaction catalysed by hexokinase can be written as follows:

$$\text{glucose} + \text{ATP} = \text{glucose 6-phosphate} + \text{ADP}$$

(ignoring complications that arise from the fact that three of the reactants exist as mixtures of different ionic states). It has an equilibrium constant at pH 7 of about 850, which may be written in accordance with the definition of ΔG^0 in the preceding section as $\Delta G^{0'} = -5.7 \log_{10}(850) = -16.7 \text{ kJ mol}^{-1}$. (We write $\Delta G^{0'}$ to indicate that as good biochemists we are more interested in what happens at pH 7 than at pH 0, the pH implied by the symbol ΔG^0 by itself, because pH 0 means $[\text{H}^+] = 1 \text{ M}$, the standard state used by chemists.)

We saw at the beginning of Section 3.11 that the $\Delta G^{0'}$ values for consecutive reactions could be added together to get the $\Delta G^{0'}$ value of the overall process, and we can, if we wish, analyse the hexokinase reaction as the difference between two hydrolytic reactions:

$$\text{ATP} + (\text{water}) = \text{ADP} + \text{inorganic phosphate}$$

$$\text{glucose 6-phosphate} + (\text{water}) = \text{glucose} + \text{inorganic phosphate}$$

and not surprisingly the $\Delta G^{0'}$ value for the hexokinase reaction is the difference between the ΔG^0 values for these two reactions: $-16.7 \text{ kJ mol}^{-1} = -30.5 - (-13.8) \text{ kJ mol}^{-1}$. (The reason for writing water in parentheses will be given at the end of this section.) This is perfectly legitimate, but it is important to realize that the hexokinase reaction does not follow this route in the cell, and even if it did there would be no reason to divide -13.8 by -30.5 and call it the 'efficiency' of the hexokinase reaction.

One way to see that this is meaningless is to realize that if we chose a different pair of irrelevant reactants instead of (water) and inorganic phosphate (which are *not* participants in the hexokinase reaction) we would get a different result. For example, using fructose and fructose 6-phosphate as irrelevant reactants,

$$\text{ATP} + \text{fructose} = \text{ADP} + \text{fructose 6-phosphate}$$

$$\text{glucose 6-phosphate} + \text{fructose} = \text{glucose} + \text{fructose 6-phosphate}$$

with $\Delta G^{0'}$ values of -14.6 and $+2.1 \text{ kJ mol}^{-1}$ respectively, the same sort of meaningless calculation would produce an 'efficiency' of $\frac{-2.1}{14.6}$, or -14%.

Another way to see that it is meaningless is to realize that the numerical result depends arbitrarily on the standard state, which is just an arbitrary convention derived from the history of the subject. If chemical thermodynamics had been developed by biochemists rather than by chemists, it is quite likely that $0.1\,\mu M$ would have been selected as the standard state for all reactants and not just for protons, and if this had been the case the $\Delta G^{0''}$ values (the double prime emphasizing the change in definition) for hydrolysis of ATP and glucose 6-phosphate would have been -70.4 and $-53.7\,\text{kJ}\,\text{mol}^{-1}$ respectively instead of -30.5 and $-13.8\,\text{kJ}\,\text{mol}^{-1}$: these have exactly the same *difference* as before ($-16.7\,\text{kJ}\,\text{mol}^{-1}$) but a different *ratio*, in accordance with the idea that differences between ΔG^{0} values are meaningful but ratios are not.

Finally, it is worth asking what a ratio of ΔG^{0} values means in terms of the corresponding equilibrium constants. Suppose that $\Delta G^{0} = -RT\ln K_{1}$ and $\Delta G_{2}^{0} = -RT\ln K_{2}$, then what meaning could be attached to an 'efficiency' E defined as $E = \Delta G_{2}^{0}/\Delta G_{1}^{0}$? Clearly, in terms of the equilibrium constants this is $E = \ln K_{2}/\ln K_{1}$, or $E\ln K_{1} = K_{2}$, so $K_{2} = K_{1}^{E}$, a dimensional absurdity.

There is yet another complication that needs to be mentioned before leaving this subject. If you examine the last example in detail you may be puzzled that the $\Delta G^{0''}$ values for hydrolysis of ATP and glucose 6-phosphate change with the definition of the standard state, given that each reaction has equal numbers of reactants on both sides of the equation, so all the changes in standard state should cancel. (If you are not puzzled, you ought to be.) The reason is that in calculations of equilibria it is not usually convenient to include the concentration of the solvent in the definition of the equilibrium constant even if it is a reactant, and there is yet another convention (not a fundamental law of chemistry) which says that the solvent is omitted from the definition. Another way of saying the same thing is that the standard state of the solvent is conventionally taken to be the actual state that exists in the system.

3.14 Redox potentials

A *redox couple* is a pair of substances such that one, the *oxidized state*, can be converted into the other, the *reduced state*, by accepting one or more electrons. For example, the $Fe^{3+}\,|\,Fe^{2+}$ couple can be written as

$$Fe^{3+} + \varepsilon^{-} \rightarrow Fe^{2+}$$

Now free electrons (ε^{-}) cannot exist in aqueous solution and so this reaction, a so-called *half-reaction*, cannot occur unless there is another half-reaction occurring simultaneously in the reverse direction. Nonetheless, certain couples occur in so many different metabolic reactions that it is useful to be able to express their oxidizing ability without explicit reference to another half-reaction. This may be done by means of the *redox potential E*, which is defined by the Nernst equation:

$$E = E^{0} + \frac{RT}{n\mathcal{F}}\ln\frac{[\text{ox}]}{[\text{red}]}$$

in which E^0 is a constant for the particular couple known as the *standard redox poten-tial*, $R = 8.314\,J\,mol^{-1}\,K^{-1}$ is the gas constant, T is the temperature in kelvin, n is the number of electrons transferred in the half-reaction (for example $n = 1$ for the $Fe^{3+}\,|\,Fe^{2+}$ couple), $\mathcal{F} = 96\,495$ coulomb mol^{-1} is a conversion factor that allows E and E^0 to be measured in volts, and [ox] and [red] are, respectively, the concentrations of the oxidized and reduced forms of the couple. (Strictly, the term is *oxidation-reduction potential* rather than redox potential, but this is rather cumbersome and I shall use the shorter and more colloquial form in this discussion.)

For the $Fe^{3+}\,|\,Fe^{2+}$ couple, $E^0 = +0.77$ volt and $n = 1$, so

$$E = 0.77 + \frac{RT}{\mathcal{F}} \ln \frac{[Fe^{3+}]}{[Fe^{2+}]} = 0.77 + 0.060 \log \frac{[Fe^{3+}]}{[Fe^{2+}]} \text{ volt}$$

where the coefficient 0.060 volt is the value of $2.303\,RT/\mathcal{F}$ at 303 K (30 °C). (Although there is a standard symbol V for the volt it will not be used in this book because it is too easily confused with more common uses of V in biochemistry. It is, in fact, little used in biochemistry, probably for this reason, though the symbol mV for the millivolt is in common use.) The form in terms of natural logarithms is just as easy to use on a modern calculator as the form at the right-hand side. Nonetheless, the latter is useful for mental calculations as it tells us that each factor of 10 in the ratio of oxidized and reduced states corresponds to a potential change of 0.060 volt. However we write it, the equation tells us that the actual potential E is determined by two separate quantities, not only by the standard potential E^0 but also by the ratio of concentrations of the two components: not surprisingly, a couple becomes more powerfully oxidizing as the proportion in the oxidized state increases.

The Nernst equation is often written with a negative sign and the ratio inverted, as follows:

$$E = E^0 - \frac{RT}{n\mathcal{F}} \ln \frac{[red]}{[ox]}$$

and some books write the standard potential on the left-hand side of the equation and the measured potential of the right-hand side, with, of course, a reversed sign:

$$E^0 = E + \frac{RT}{n\mathcal{F}} \ln \frac{[red]}{[ox]}$$

All of this can seem very confusing (does the equation contain a positive or a negative sign? is [ox] in the numerator or in the denominator? on which side of the equation should the standard potential appear, etc.), but actually it is not difficult to avoid confusion. First of all, it should be evident from the discussion of the properties of logarithms earlier in this chapter that these last two forms (as well as another that you can easily deduce for yourself) are exactly equivalent to the one given first, because changing the sign of a logarithm is equivalent to taking the reciprocal of the value whose logarithm it is, as noted at the end of Section 3.5. Second, it is easy to remember that whichever way the equation is written increasing the proportion of *reduced* form *reduces* the value of the measured potential.

The meaning of the standard potential E^0 can be understood by putting [red] = [ox] = 1 M. (Actually the particular concentration is not important for the mathematical argument, which would be unchanged if [red] = [ox] at some concentration other than 1 M. However, this concentration is specified for defining the standard states and in practice it ought to be specified because the Nernst equation may not be obeyed exactly.) Then the logarithmic term becomes zero (the logarithm of 1 is 0 regardless of base, and regardless of whether we multiply the result by some factor) and so

$$E = E^0$$

in these conditions. Thus the standard potential E^0 is the value the actual potential E has when the components of the couple are in their standard states.

3.15 **Standard states**

Mention of standard states at the end of the last section brings us to a point that is actually quite simple if one thinks about it with a clear head even though it often generates confusion. In the hopes of making it seem as simple as it really is, let us discuss it in terms of distances between places rather than in terms of redox potentials. In any road atlas one can find charts of the kind illustrated in Table 3.2(a), which show the distances between selected places, for example, if we wanted to know the distance from Rugby to St Albans we could read it off the chart as 98 km. However, anyone familiar with the route from Birmingham to London will realize that this example makes it more complicated than it need be because all the places listed are along the same route. So we could provide all of the information with a much simpler list that just showed the distances from London, as in Table 3.2(b): $E(\text{London}) = 0$ km, $E(\text{St Albans}) = 21$ km, and so on. From such a list we can get any desired distance by subtracting the corresponding distances from London, for example the distance from Rugby to St Albans is $119 - 21 = 98$ km, exactly as for the two-dimensional chart.

This works in this example because we have used a one-dimensional representation of a one-dimensional problem. It would not work if the problem were genuinely two-dimensional. For example, we could not include Bristol in the simple list and expect to get the correct distance from Birmingham by comparing the distances from Birmingham to London and from London to Bristol, because most people do not go from Birmingham to Bristol via London. That is why distance charts in road atlases are nearly always two-dimensional: one-dimensional charts would give hopelessly inaccurate results except in countries like Canada or Chile where the main cities are strung out (to a first approximation) in a long line. However, the problem of redox potentials is one-dimensional, so we can stay with

Table 3.2 Distance charts: (**a**) a two-dimensional chart; (**b**) a one-dimensional chart with London defined as the starting point; (**c**) a one-dimensional chart with Dunstable defined as the starting point.

(a)							(b)	(c)
Birmingham	0						164	124
Coventry	29	0					135	95
Rugby	45	16	0				119	79
Newport Pagnell	95	66	50	0			69	29
Dunstable	124	95	79	29	0		40	0
St Albans	143	114	98	48	19	0	21	−19
London	164	135	119	69	40	21	0	0	−40
	B	C	R	NP	D	StA	L	

the example illustrated in Table 3.2. (The reason *why* it is one-dimensional comes from thermodynamics rather than mathematics, so we do not need to discuss in it this book, but it has to do with the fact that it is impossible to extract work from a cyclic system at equilibrium).

Although we measured the one-dimensional distances from London, we were not obliged to do that: they could equally well be measured from any place along the route or its extension beyond London or beyond Birmingham. The only proviso would be that if one chose an intermediate point, say Dunstable, one would have to take care to give opposite signs for the distances in the two directions, for example positive distances in the direction of Birmingham, negative ones in the direction of London. Although this choice, which is illustrated in Table 3.2(c), would seem to be a little odd to anyone who did not live in Dunstable, it would give perfectly correct results. For example, the distance from Rugby to St Albans is now calculated as $79 - (-19)$ km, but the result is still 98 km, exactly as before.

The point of all this is that when distances are measured from an arbitrary starting point the individual values are arbitrary, but the *differences* between them are not arbitrary. It only makes sense to say that Dunstable is about twice as far as St Albans if 'far' has previously been defined to mean 'far from London', but it makes sense to say that Dunstable is 19 km from St Albans without needing to specify the starting point. Put more generally, quantities that are measured from an arbitrary standard state can always be compared as *differences*, but should only be compared as *ratios* in very carefully defined circumstances.

Returning now to the redox potentials, the fundamental experimental problem is that we cannot study a half-reaction in isolation: we cannot, for example, look at the $Fe^{3+} + \varepsilon^- \rightarrow Fe^{2+}$ reaction all by itself. The best we can do is compare it with another reaction, such as $Cu^{2+} + \varepsilon^- \rightarrow Cu^+$ (chosen so that the unobservable species ε^- cancels from the net reaction) and look at the difference between the oxidizing potentials of the two half-reactions. When I was taught chemistry I was taught that the value of a redox potential had an absolute physical meaning but one could not measure it; I now think that is wrong and confusing: the truth seems to be that a redox potential has no absolute physical meaning, any more than it has meaning to say 'Dunstable has a distance of 79 km'.

To allow E^0 values to be calculated from one another and to avoid having to set up a huge two-dimensional table including comparisons between all known half-reactions, it is far more convenient to *define* one particular E^0 value as zero and measure all others as differences from it. The conventional choice for this purpose is *the standard hydrogen electrode* $H^+ | \frac{1}{2} H_2$ in which dissolved protons are maintained in equilibrium with hydrogen gas under standard conditions.

The $Fe^{3+} | Fe^{2+}$ couple is particularly simple as an introduction to redox potentials because no protons are involved in the half-reaction and, in consequence, its potential is independent of the hydrogen-ion concentration. Thus, even though E^0 is defined at pH 0, it would have exactly the same value of 0.77 volt for the $Fe^{3+} | Fe^{2+}$ couple if it were defined at pH 7 or some other pH. This is not true for most couples of biochemical interest, because most involve a transfer of protons as well as electrons. The simplest case of a pH-dependent couple occurs when none of the reactants ionizes in the pH range considered, and the protons transferred can then just be included in the Nernst equation as reactants. For example, two protons are transferred by the acetaldehyde | ethanol couple:

$$CH_3CHO + 2H^+ + 2\varepsilon^- \rightarrow C_2H_5OH$$

and has $E^0 = 0.221$ volt. So the potential is given by

$$E = 0.221 + \frac{RT}{2\mathcal{F}} \ln \frac{[CH_3CHO][H^+]^2}{[C_2H_5OH]}$$

$$= 0.221 + \frac{RT}{2\mathcal{F}} \ln \frac{[CH_3CHO]}{[C_2H_5OH]} + \frac{RT}{2\mathcal{F}} \ln [H^+]^2$$

$$= 0.221 + \frac{RT}{2\mathcal{F}} \ln \frac{[CH_3CHO]}{[C_2H_5OH]} + \frac{RT}{\mathcal{F}} \ln [H^+]$$

$$= 0.221 + \frac{RT}{2\mathcal{F}} \ln \frac{[CH_3CHO]}{[C_2H_5OH]} + \frac{2.303RT}{\mathcal{F}} \log[H^+]$$

$$= 0.221 + \frac{RT}{2\mathcal{F}} \ln \frac{[CH_3CHO]}{[C_2H_5OH]} - \frac{2.303RT}{\mathcal{F}} pH$$

$$= 0.221 + \frac{RT}{2\mathcal{F}} \ln \frac{[CH_3CHO]}{[C_2H_5OH]} - 0.060 \, pH \quad volt$$

The first form of this expression is just what we get if we substitute $n = 2$, $[ox] = [CH_3CHO][H^+]^2$ (since $ox = CH_3CHO + H^+ + H^+$ and, therefore, $[ox] = [CH_3CHO][H^+][H^+]$) and $[red] = [C_2H_5OH]$ into the Nernst equation. The others follow from various properties of logarithms that we have considered already in this chapter. First of all, the fact that $\log xy = \log x + \log y$ allows the proton term to be written separately from the term for the other reactants. Next, the fact that $\ln x^2 = 2 \ln x$ allows the 2 in the denominator of the coefficient to be cancelled with the 2 as the exponent. Then the conversion factor 2.303 ($= \ln 10$) allows us to convert from natural to common logarithms, and the definition of pH allows $-pH$ to be written instead of $\log[H^+]$. Finally, as $2.303 RT/\mathcal{F}$ is a constant of known value, 0.060 volt, we can introduce this value in the last line. Thus, at any $[CH_3CHO]/[C_2H_5OH]$ ratio the potential decreases by 0.06 volt for each unit increase in the pH and is 0.42 volt less at pH 7 than at pH 0.

Of the couples of biochemical importance, some, for example cytochrome $c(Fe^{3+})$|cytochrome $c(Fe^{2+})$, resemble Fe^{3+}|Fe^{2+} in having redox potentials that do not vary with pH; a few, such as acetaldehyde|ethanol, show a decrease of 0.06 volt for each unit increase in pH, but many show intermediate behaviour. This is because we can assume neither that protons are not involved (as for Fe^{3+}|Fe^{2+}), nor that the reactants remain in the same ionic state of the pH range of interest (as for acetaldehyde|ethanol). Instead, we have to allow for the complication that the half-reaction is not the same process at every pH. We cannot just ignore this complication because it applies to the majority of half-reactions of biochemical interest; the simple cases we have considered are the exceptions, so we shall examine the more common case in the next section. Meanwhile, we must realize that naively applied E^0 values are a very misleading guide to redox behaviour at neutral pH values. For example, cytochrome $c(Fe^{3+})$|cytochrome $c(Fe^{2+})$ and acetaldehyde|ethanol have fairly similar E^0 values, 0.250 and 0.221 volt, respectively, but it would be quite wrong to think that the former would have only a weak capacity to oxidize the latter under metabolic conditions: on going from pH 0 to pH 7, the former potential remains unchanged at 0.250 volt, whereas the latter decreases to -0.199 volt. To decrease this confusion, biochemists commonly *redefine* the standard as the observed potential at a specified pH (usually pH 7) when the two components are at equal concentrations. This potential, often called the *mid-point potential*, is given the symbol $E^{0'}$, and the Nernst equation becomes

$$E = E^{0'} + \frac{RT}{n\mathcal{F}} \ln \frac{[ox]}{[red]}$$

Proton concentrations are not included in this equation and so it is valid only at the pH at which $E^{0'}$ is defined. From the calculation just done, we have $E^{0'} = -0.199$ volt for acetaldehyde|ethanol at pH 7, but still $E^{0'} = E^0 = +0.250$ volt for cytochrome $c(Fe^{3+})$|cytochrome $c(Fe^{2+})$.

If two redox couples are allowed to interact, whether by being mixed together in the presence of a suitable catalyst or by connecting them electrically, they can react stoichiometrically until their potentials are identical. For example, in the reaction catalysed by alcohol dehydrogenase the acetaldehyde|ethanol couple reacts with the $NAD_{ox}|NAD_{red}$ couple:*

$$NAD_{ox} + 2\varepsilon \rightarrow NAD_{red}, \quad E^{0'} = -0.320 \text{ volt (pH 7, 30°C)}$$

Each of the two interacting couples has its own potential defined by its own Nernst equation, and when they are equal, i.e. at equilibrium, we can equate the two right-hand sides:

$$-0.199 + 0.03 \log\frac{[CH_3CHO]}{[C_2H_5OH]} = -0.320 + 0.03 \log\frac{[NAD_{ox}]}{[NADH_{red}]}$$

Rearranging this, we have

$$0.03\left(\log\frac{[NAD_{ox}]}{[NADH_{red}]} - \log\frac{[CH_3CHO]}{[C_2H_5OH]}\right) = 0.320 - 0.199 = 0.121 \text{ volt}$$

Dividing both sides by 0.03 volt and applying the rule for subtracting logarithms gives

$$\log\frac{[NAD_{ox}][C_2H_5OH]}{[NADH_{red}][CH_3CHO]} = \frac{0.121}{0.03} = 4.033$$

and as antilog $4.033 = 1.08 \times 10^4$, 4.033 can be written as $\log(1.08 \times 10^4)$, and the whole equation can be written without the log function:

$$\frac{[NAD_{ox}][C_2H_5OH]}{[NADH_{red}][CH_3CHO]} = 1.08 \times 10^4$$

which is the equilibrium constant for the reaction at pH 7 and 30°C.

Like the standard Gibbs energy of reaction $\Delta G^{0'}$, the difference between two standard redox potentials is a logarithmic expression of an equilibrium constant, and in general we can express this difference $\Delta E^{0'}$ as follows:

$$\Delta E^{0'} = \frac{-RT}{n\mathcal{F}} \ln K$$

---

* In accordance with recent recommendations of the IUBMB, charges and protons are not written in equations that do not refer to defined ionic species, i.e. equations that refer to a mixture of ionic states at a specified pH. For this reason the protons are omitted along with the charges on the electron and the two forms of NAD. Although the symbols NAD^+ and NADH are very common in the literature instead of NAD_{ox} and NAD_{red}, they are not really satisfactory as *both* of them are negatively charged at neutral pH. Even though 'everybody knows' that the + in NAD^+ refers to the nicotine ring and not to the whole molecule, it is poor practice to use + as the label for an anion.

whereas

$$\Delta G^{0'} = -RT \ln K$$

and, so dividing one expression by the other:

$$\Delta E^{0'} = \frac{-RT}{n\mathcal{F}} \ln K$$

Thus, standard redox potential differences and standard Gibbs energy of reaction are measurements of the same thing but in different units, volts for $\Delta E^{0'}$ and $kJ \, mol^{-1}$ for $\Delta G^{0'}$. Redox potentials are commonly given in electrical units because they are often *measured* in electrical units, electrochemical cells providing the most convenient and accurate way of determining equilibrium constants accurately, especially for reactions in which the equilibrium position is very far in one direction.

Given that Gibbs energies and differences in potential are the same thing, it would be convenient if a single convention were adopted and they were always expressed in the same units but, unfortunately, biochemistry has not developed in that way: equilibrium data for complete reactions are nearly always given as $\Delta G^{0'}$ values in $kJ \, mol^{-1}$, whereas data for half-reactions are commonly expressed as $E^{0'}$ values in volts (or millivolts). This makes it quite difficult for the general biochemist to make easy sense of the specialized literature on electron-transfer reactions, which is full of values expressed in millivolts (mV). However, as Table 3.3 illustrates, the relevant numbers are not so difficult to memorize, especially if, as in the table, there is no attempt at high accuracy. Every $6 \, kJ \, mol^{-1}$ in a $\Delta G^{0'}$ value corresponds to about a factor of 10 in the equilibrium constant, and to about $60 \, mV$ in the difference in mid-point potentials of the two half-reactions (for a one-electron transfer). In other words, $1 \, kJ \, mol^{-1}$ is about $10 \, mV$. As shown in the table, the relationships are different if more than one electron is transferred, but this is not a major problem because in biochemical practice one mostly sees values given in mV for cytochromes, iron–sulphur proteins, haem-proteins, etc., all of which are involved in one-electron transfers.

Table 3.3 Correspondence between equilibrium constants, standard Gibbs energies, and mid-point potentials.

K	$\Delta G^{0'}$ ($kJ \, mol^{-1}$)	$\Delta E^{0'}(n=1)$ (mV)	$\Delta E^{0'}(n=2)$ (mV)
0.001	18	180	90
0.01	12	120	60
0.1	6	60	30
1	0	0	0
10	−6	−60	−30
100	−12	−120	−60
1000	−18	−180	−90

3.16 Dependence of redox potentials on pH

For many redox couples one or sometimes both of the components ionize between pH 0 and pH 7 and consequently E varies with pH in a more interesting way than in either of the simple examples of pH behaviour discussed in the preceding section. Consider, for example, the couple acetic acid|acetaldehyde:

$$CH_3CO_2H + 2H^+ + 2\varepsilon^- \rightarrow CH_3CHO + H_2O$$

for which $E^0 = -0.11$ volt. If we made a false analogy with the acetaldehyde|ethanol couple considered before, we might expect the potential at pH 7 to be 0.42 volt smaller, i.e. -0.53 volt, but this would ignore the fact that the oxidized form acetic acid ionizes in the pH range considered, with a pK_a $(= -\log K_a)$ of 4.73. Thus, at pH 7 the predominant half-reaction is not as shown above, but

$$CH_3CO_2^- + 3H^+ + 2\varepsilon^- \rightarrow CH_3CHO + H_2O$$

Box 3.3 How to handle the variation of a redox potential with pH

1. The Nernst equation at pH 0 is normally expressed in terms of a *single* oxidizing form (e.g. CH_3CO_2H) and a *single* reducing form (e.g. CH_3CHO):

$$E = E^{0'} + \frac{RT}{n\mathcal{F}} \ln \frac{[CH_3CO_2H]}{[CH_3CHO]}$$

2. In many couples of biochemical importance one or both forms ionize between pH 0 and pH 7: e.g. CH_3CO_2H is almost fully protonated at pH 0, but almost fully deprotonated ($CH_3CO_2^-$) at pH 7; CH_3CHO does not ionize in this range.

3. Experimentally, the *total* concentrations of oxidized ($[CH_3CO_2(H)] = [CH_3CO_2H] + [CH_3CO_2^-]$) and reduced forms are ordinarily controlled, not the concentrations of any particular ionic states.

4. The Nernst equation therefore needs to be re-expressed in terms of total concentrations.

5. This can be done by substituting the usual equation for an ionic equilibrium,

$$[CH_3CO_2H] = \frac{[CH_3CO_2(H)]}{1 + K_a/[H^+]}$$

into the original expression of the Nernst equation:

$$E = E^{0'} + \frac{RT}{n\mathcal{F}} \ln \frac{[CH_3CO_2(H)]}{(1 + K_a/[H^+])[CH_3CHO]}$$

As mentioned at the end of Section 3.13, the usual convention is to define the standard state of the solvent as the state that exists in the system. This allows water to be omitted when writing the Nernst equation for the couple:

$$E = E^0 + \frac{RT}{2\mathcal{F}} \ln \frac{[CH_3CO_2H]}{[CH_3CHO]} + \frac{RT}{\mathcal{F}} \ln [H^+]$$

This form is exactly analogous to that written for the acetaldehyde|ethanol couple when the proton term was written separately.

The equation is valid at any pH, even when acetic acid exists almost entirely as acetate ion, but 'valid' does not mean 'convenient', and in fact this form of the equation would be very inconvenient to use at high pH, because it is difficult to maintain the CH_3CO_2H concentration constant as the pH changes. In reality, we would usually keep constant the *total* concentration of all ionic states of the oxidized form, i.e. $[CH_3CO_2(H)] = [CH_3CO_2H] + [CH_3CO_2^-]$; so it would be much more helpful to recast the Nernst equation in terms of this total concentration, and we get an expression for this by examining the definition of the acid dissociation constant K_a:

$$K_a = \frac{[CH_3CO_2^-][H^+]}{[CH_3CO_2H]}$$

Hence

$$[CH_3CO_2^-] = [CH_3CO_2H]\,\frac{K_a}{[H^+]}$$

So

$$[CH_3CO_2(H)] = [CH_3CO_2H] + [CH_3CO_2^-]$$

$$= [CH_3CO_2H]\left(1 + \frac{K_a}{[H^+]}\right)$$

and thus the concentration of acid is readily expressed in terms of the total concentration and the concentration of hydrogen ions:

$$[CH_3CO_2H] = \frac{[CH_3CO_2(H)]}{1 + \dfrac{K_a}{[H^+]}}$$

Substituting this for $[CH_3CO_2H]$ in the Nernst equation, we get

$$E = E^0 + \frac{RT}{2\mathcal{F}} \ln \frac{[CH_3CO_2(H)]}{[CH_3CHO]\left(1 + \dfrac{K_a}{[H^+]}\right)} + \frac{RT}{\mathcal{F}} \ln [H^+]$$

This looks rather complicated, but remember that the logarithm of any product can be written as a sum or difference of separate terms, so this one can be written as follows:

$$E = E^0 + \frac{RT}{2\mathcal{F}} \ln \frac{[CH_3CO_2(H)]}{[CH_3CHO]} - \frac{RT}{2\mathcal{F}} \ln \left(1 + \frac{K_a}{[H^+]}\right) + \frac{RT}{\mathcal{F}} \ln [H^+]$$

The mid-point potential at any pH is the potential when the (total) concentrations of oxidized and reduced forms are equal, i.e. $[CH_3CO_2(H)] = [CH_3CHO]$, and as the logarithm of 1 is zero the second term disappears in this case and the expression for the mid-point potential is as follows:

$$E^{0'} = E^0 - \frac{RT}{2\mathcal{F}} \ln \left(1 + \frac{K_a}{[H^+]}\right) + \frac{RT}{\mathcal{F}} \ln [H^+]$$

or in numerical terms:

$$E^{0'} = -0.11 - 0.03 \log \left(1 + \frac{K_a}{[H^+]}\right) - 0.06 \, pH \quad volt$$

We can now deduce in general terms what this expression predicts by considering how it simplifies when $[H^+]$ is very large or very small. When $[H^+]$ is much larger than K_a, i.e. at low pH, $K_a/[H^+]$ is a very small number, and so $1 + K_a/[H^+]$ is hardly any different from 1, which has a logarithm of 1; so,

$$E^{0'} \approx -0.11 - 0.06 \, pH \quad volt \quad at \, low \, pH$$

On the other hand, when $[H^+]$ is much smaller than K_a, i.e. at high pH, $K_a/[H^+]$ is a very large number, and so $1 + K_a/[H^+]$ is hardly any different from $K_a/[H^+]$; so,

$$E^{0'} \approx -0.11 - 0.03 \log \left(\frac{K_a}{[H^+]}\right) - 0.06 \, pH$$

$$= -0.11 - 0.03 \log K_a + 0.03 \log [H^+] - 0.06 \, pH$$

$$= -0.11 - 0.03 \, pK_a - 0.09 \, pH$$

$$= -0.11 + 0.14 - 0.09 \, pH$$

$$= +0.03 - 0.09 \, pH \quad volt \quad at \, high \, pH$$

(The fact that 0.03 occurs both at the beginning and at the end of these lines is a coincidence: the first 0.03 is the value of $RT/n\mathcal{F}$; the second is the result of adding 0.14 to -0.11.)

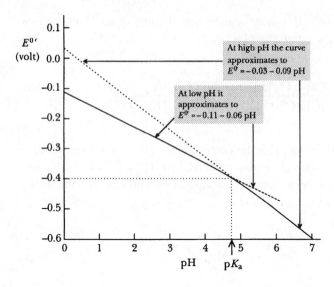

Fig. 3.2 Dependence of the mid-point potential on pH for the acetic acid | acetaldehyde couple.

These two equations tell us the limiting behaviour at the extremes, but what of the intermediate region where there is presumably a transition between the two? If $[H^+] = K_a$, then pH is the same as pK_a, and $1 + (K_a/[H^+]) = 1 + 1 = 2$; so, the general equation for the mid-point potential takes the particular form

$$E^{0'} = -0.11 - 0.03 \log 2 - 0.06\, pK_a = -0.40\, \text{volt} \quad (\text{when pH} = pK_a)$$

From these three expressions it is clear that the mid-point potential depends on pH as shown in Fig. 3.2: at low pH, $\Delta E^{0'}$ decreases linearly by 0.06 volt for each unit increase in pH; at high pH it decreases linearly by 0.09 volt for each unit increase in pH; the two linear regions are joined by a curve, but the extrapolated straight lines intersect at pH = pKa.

....

Example 3.2 Calculation of a potential

The couple gluconic acid | glucose refers to the following process:

$$\text{gluconic acid} + H^+ + 2\varepsilon^- \rightarrow \text{glucose} + H_2O$$

and has a standard redox potential of -0.73 volt. Glucose does not ionize in the pH range 0–7 but gluconic acid has a pK_a of 3.4. Calculate the mid-point potential at pH 7.

The pK_a of 3.4 implies an acid dissociation constant of $10^{-3.4}$, or antilog$(-3.4) = 2.5 \times 10^{-4}$, and writing gluconic acid as GH, gluconate as G^- and the sum of their concentrations as $[G]_{total} = [GH] + [G^-]$, this gives

$$[G^-][H^+] = 2.5 \times 10^{-4}[GH]$$

Hence

$$[G]_{total} = [GH](1 + 2.5 \times 10^{-4}/[H^+])$$

or

$$[GH] = [G]_{total}/(1 + 2.5 \times 10^{-4}/[H^+])$$

which can be substituted into the expression for the redox potential:

$$E = -0.73 + \frac{RT}{2\mathcal{F}} \ln \frac{[GH][H^+]}{[Glc]}$$

$$= -0.73 + \frac{RT}{2\mathcal{F}} \ln \frac{[G]_{total}[H^+]}{[Glc](1 + 2.5 \times 10^{-4}/[H^+])}$$

At pH 7.0, $[H^+] = 10^{-7}$ and, at the mid-point, $[G]_{total} = [Glc]$, so this simplifies to

$$E = -0.73 + \frac{RT}{2\mathcal{F}} \ln \frac{10^{-7}}{1 + 2.5 \times 10^{-4}/10^{-7}} = -0.45 \text{ volt}$$

I have discussed redox potentials and their dependence on pH in rather more detail than one would usually expect in an elementary mathematics book because it is a topic that causes a great deal of misunderstanding and muddle, not only among students but also, regrettably, among the authors of biochemistry text-books. Most of this misunderstanding can be avoided by studying the simple logic reviewed in **Box 3.3**. By following the rules set out there, it is quite straight-forward to deal with couples that involve more than a single ionization. For example, the couple pyruvic acid|lactic acid,

$$CH_3COCO_2H + 2H^+ + 2\varepsilon^- \rightarrow CH_3CHOHCO_2H$$

has $E^0 = 0.29$ volt, $pK_{a(pyr)} = 2.50$, $pK_{a(lact)} = 3.86$ and thus requires ionizations of both oxidized and reduced forms to be taken into account. Nonetheless, a

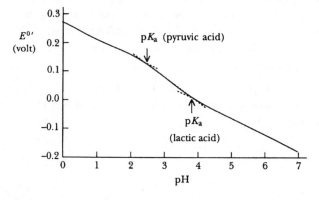

Fig. 3.3 Dependence of the mid-point potential on pH for the pyruvic acid|lactic acid couple.

derivation along the lines of that shown above for acetic acid|acetaldehyde is not difficult and leads to

$$E^{0\prime} = 0.29 - 0.03 \log\left(\frac{1 + K_{a(pyr)}}{[H^+]}\right) + 0.03 \log\left(1 + \frac{K_{a(lact)}}{[H^+]}\right) - 0.06 \, pH \quad \text{volt}$$

and the dependence on pH shown in Fig. 3.3.

3.17 Problems

3.1 Evaluate the following expressions:

(a) 3^3,　　　(b) 4^{-2},　　(c) $27^{1/3}$

(d) $4^{-1/2}$,　　(e) 16^0,　　(f) 10^{-3}

(g) $3.29^6 \times 3.29^{-4} \times 3.29^{-3} \times 3.29$,　　(h) $\overline{2}.37$

(i) $4!$,　　　　(j) antilog$_3$3,　　(k) log 100

(l) $\log_2 8$,　　(m) antilog 3

3.2 Without using a calculator or tables, write down approximate values of

(a) $e^{0.16}$,　　(b) $\exp(-0.077)$

3.3 (a) Rearrange the approximate formula you used in problem (3.2) to give an approximate formula for $\ln(1 + x)$ that applies when x (positive or negative) is numerically small compared to 1. Then use it to write down approximate values of

(b) $\ln 1.116$,　　(c) $\ln 0.983$

(d) $\ln 1.041$,　　(e) $\ln 0.888$

3.4 Given that $\ln 2 = 0.693$, $\ln 3 = 1.099$, $\ln 5 = 1.609$, $\ln 7 = 1.946$, evaluate the following (without using the ln or exp keys on a calculator):

(a) $\ln 4$,　　(b) $\ln 0.2$,　　(c) $\ln 27$

(d) $\ln 0.6$,　　(e) $\ln 1$,　　(f) $\exp(-0.693)$

(g) $\exp(1.099)$,　　(h) $\exp(2.708)$,　　(i) $\exp(0.337)$

3.5 According to the Boltzmann principle, the number of molecules at thermal equilibrium in a state with energy E is proportional to $\exp(-E/RT)$, where $R = 8.31 \, J \, mol^{-1} K^{-1}$ is the gas constant and T is the temperature in kelvin. In a proton magnetic resonance experiment, any nucleus can exist in either of two spin states differing in energy (in a 100 MHz instrument) by about $4 \times 10^{-2} J \, mol^{-1}$. Calculate the relative populations of nuclei in the two states.

3.6 Sketch the form of a plot of mid-point potential against pH for a couple of the form $Ox + 2H^+ + 2\varepsilon^- \rightarrow RedH$, assuming that $E^0 = +0.100$ volt and RedH is a monobasic acid with $pK_a = 4.3$.

CHAPTER 4

Differential calculus

4.1 Coordinate geometry

Traditional geometry is concerned with shapes, on paper or in space, and with the relationships between shapes. It has two distinct extensions into the more numerical domains of mathematics: *trigonometry and coordinate geometry*. Trigonometry is concerned with calculating the relationships between lengths and angles, whereas coordinate geometry has almost the reverse objective: it allows the use of geometrical insight and understanding for studying problems that are not essentially geometrical at all but algebraic. It happens that biochemistry is rather little concerned with real shapes, distances or angles; for the biochemist, therefore, neither traditional geometry nor trigonometry occupies the centre of the mathematical stage. Coordinate geometry, by contrast, is crucial, because many important relationships in physical chemistry and biochemistry appear in the first instance as algebraic expressions, and in this form they are too abstract to be immediately understood. The biochemist thus has a frequent need to draw graphs to represent mathematical relationships as lines on paper (**Box 4.1**).

As an introduction to the use of coordinate geometry, let us consider the following simple equation:

$$y = 7 + 3x$$

which expresses a relationship between two variables x and y. The variable y on the left-hand side is expressed as a function of the variable x on the right-hand side, and although sometimes we may want to write the equation the other way round (see Section 6.2), there is often an implied 'direction' for the relationship, and when we plot one against the other it is usual to plot the left-hand variable

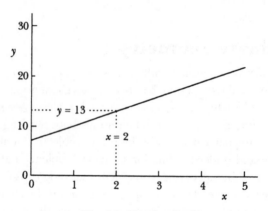

Fig. 4.1 Plot of $y = 7 + 3x$.

on the vertical axis, known as the *ordinate* and the right-hand variable on the horizontal axis, known as the *abscissa*.

When we plot y against x as shown in Fig. 4.1 we are in effect using a geometrical model to 'map' an algebraic equation, giving concrete expression to what would otherwise be an abstraction. For example, if we let $x = 2$ the equation tells us that $y = 7 + 3 \times 2 = 13$, and we can plot the relationship by marking a point 2 units from the y-axis in the x direction and 13 units from the x-axis in the y direction. We can refer to this point by means of the shorthand (2, 13), which just means 'the point where $x = 2$ and $y = 13$'. It turns out that all points defined by the equation lie on a straight line, as shown in Fig. 4.1, and indeed any equation of the form

$$y = a + bx$$

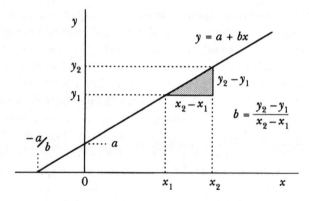

Fig. 4.2 Plot of the general straight line $y = a + bx$.

where a and b are constants, defines a straight line if y is plotted against x. This more general straight line is shown in Fig. 4.2. In elementary mathematics, a and b are frequently replaced by c and m, respectively, and the two terms on the right-hand side are often written in the opposite order, i.e. the same equation is often written

$$y = mx + c$$

This is fine as long as we remember that the symbols do not have any *universal* meaning of the sort implied here. When we come to generalize this sort of relationship to more complicated cases, it is usually more convenient to order the terms from lowest- to highest-order, and to use a notation that makes it more obvious what comes next than it is with $c, m, \ldots$

If we put $x = 0$, then $bx = 0$ also, and hence a is the value of y when $x = 0$. The straight line thus cuts the y-axis at a distance a from the point where the axes intersect, which is known as the *origin*, and has the coordinates $(0, 0)$. Because of this relationship, a is known as the *intercept* on the y-axis. The intercept on the x-axis is less obvious, but it may readily be found by putting $y = 0$: this gives $a + bx = 0$, and so $x = -b/a$, which is therefore the intercept on the x-axis. Notice that there are intercepts on both axes, but in everyday use if the axis is not specified 'the intercept' usually means the intercept on the y-axis.

The meaning of b in the general equation for a straight line follows from consideration of how y changes when x changes. Suppose that y changes from y_1 to y_2 as x changes from x_1 to x_2. Then

$$y_1 = a + bx_1, \qquad y_2 = a + bx_2$$

If the first equation is subtracted from the second, the constant a disappears:

$$y_2 - y_1 = a + bx_2 - a - bx_1 = bx_2 - bx_1 = b(x_2 - x_1)$$

and if we divide both sides of the equation by $(x_2 - x_1)$ and interchange the left- and right-hand sides we obtain the expression for b:

$$b = \frac{y_2 - y_1}{x_2 - x_1}$$

Notice that this relationship applies for any values of x_1 and x_2 we could have chosen (other than equal values, $x_1 = x_2$, which would introduce complications that I prefer to avoid at present). In other words, as b is constant the ratio $(y_2 - y_1)/(x_2 - x_1)$ is also constant, and equal to b; it is just this constancy that defines algebraically what we mean when we say that a line is straight.

If b is numerically very small, the straight line defined by the equation $y = a + bx$ is almost parallel with the x-axis: y changes very slowly as x changes. Conversely, if b is numerically very large, y changes very rapidly as x changes. Thus, b has a meaning very similar to that of the *gradient* of a hill in everyday life: the steeper the hill the greater the gradient, and the more rapidly the height increases with the horizontal distance travelled. The mathematical concept represented by b is also sometimes called the gradient of the line, but the word most commonly found in scientific terminology is *slope*. As we shall see in the rest of this chapter, not only straight lines have slopes but we can also usefully extend the term to apply it to any curve, and it is this extension that is the concern of differential calculus.

4.2 Slope of a curve

The curve shown in Fig. 4.3 is defined by the following equation:

$$y = 2 + 3x + x^2$$

The curve is known as a *parabola*, but this is not important for our present purpose. We saw that for the straight line considered in the previous section the ratio $(y_2 - y_1)/(x_2 - x_1)$ was a constant, and we introduced the term *slope* for it. Let us now examine the same ratio in the more complex example given in Fig. 4.3.

First, it is convenient to introduce a new notation for $(y_2 - y_1)$ and $(x_2 - x_1)$, which are often called *increments* in y and x, respectively. (An increment is similar to an increase, except that the word 'increase' normally implies a positive change

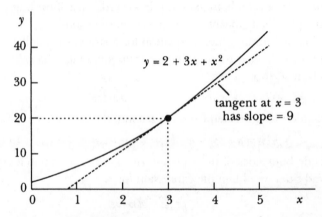

Fig. 4.3 Plot of the parabola $y = 2 + 3x + x^2$.

Table 4.1 Symbols for a change $(x_2 - x_1)$ in a variable x.

Symbol	Spoken as	Meaning
Δx	delta x	Change of any magnitude, large or small
δx	delta x	Small change
dx	d x	Infinitesimal change (approaching zero)
∂x	partial d x	Infinitesimal change under specified conditions

The prefixes in the first two lines are both the greek letter delta. As they are not usually used together in the same context it is not usually necessary to distinguish them in speech, but when a distinction is necessary Δ is a capital delta and δ is a lower-case delta. The symbol in the fourth line is called a 'curly d' (describing its shape) or a 'partial d' (describing its use).

whereas the word 'increment' carries no such implication: an increment may be positive or negative, but not zero.) Four different ways of denoting a change in x such as $(x_2 - x_1)$ are in common use, which are rather similar in appearance but distinct in meaning. Although we only need one of these immediately, they are all listed in Table 4.1 for future reference. If Δx represents a change in x, then Δy represents the *corresponding* (not just any) change in y, i.e. the change in y that occurs as a result of the change in x, either experimentally or because of the mathematical relationship between x and y. So we can write

$$y = 2 + 3x + x^2$$

$$y + \Delta y = 2 + 3(x + \Delta x) + (x + \Delta x)^2$$
$$= 2 + 3x + 3\Delta x + x^2 + 2x\Delta x + (\Delta x)^2$$

Subtracting the first equation from the second gives an expression for Δy:

$$\Delta y = 3\Delta x + 2x\Delta x + (\Delta x)^2$$
$$= \Delta x(3 + 2x + \Delta x)$$

and division of both sides of the equation by Δx produces an expression for the ratio of changes:

$$\frac{\Delta y}{\Delta x} = 3 + 2x + \Delta x$$

As we have said nothing about *how large* a change in x we are discussing, we have no grounds for simplifying this expression, but suppose we specify that Δx is a *small* change in x. This immediately raises the question of how small is small? In the present context it means small compared with $3 + 2x$, and let us emphasize this by replacing Δx with δx (a lower-case delta instead of a capital, as in Table 4.1). Then, let us watch the value of

$$\frac{\delta y}{\delta x} = 3 + 2x + \delta x$$

as we make δx smaller and smaller, as illustrated in Table 4.2 for the numerical value $x = 3$ and decreasing values of δx. Not surprisingly, δy gets smaller and

Table 4.2 Values of the increment ratio.

δx	$\delta y = \delta x(9 + \delta x)$	$\dfrac{\delta y}{\delta x}$
1	10	10
0.5	4.75	9.5
0.2	1.84	9.2
0.1	0.91	9.1
0.01	0.090 1	9.01
0.001	0.009 001	9.001
0.000 1	0.000 900 01	9.000 1
0.000 01	0.000 090 00	9.000 01
0.000 001	0.000 009 00	9.000 001
0.000 000 1	0.000 000 90	9.000 000 1

The equation $y = 2 + 3x + x^2$ is used as an example, with $x = 3$ and decreasing values of δx.

smaller as well, so we might think that the ratio would become indefinite, given that the fraction $0/0$ is not allowed in mathematics and is said to give an indefinite result (Section 1.6). However, if we follow the numbers in the right-hand column, we see that far from becoming vague or indefinite the ratio gets closer and closer to the perfectly definite number 9. It is not difficult to believe that this would continue, so that it would approach a value of 9 when δx became vanishingly small.

We can express this by saying that the *limit* of $\dfrac{\delta y}{\delta x}$ as δx approaches zero is $3 + 2x$, or

$$\lim_{\delta x \to 0} \frac{\delta y}{\delta x} = 3 + 2x$$

As illustrated in Fig. 4.3, this result gives the slope of the *tangent* to the curve $y = 2 + 3x + x^2$ at the point where the expression is evaluated. For convenience, we often refer to this simply as the slope of the curve itself, although unlike the slope of a straight line it is not a constant.

Expressions such as $\lim_{\delta x \to 0} \dfrac{\delta y}{\delta x}$ are much too cumbersome for common use and so we *define* a new quantity $\dfrac{dy}{dx}$ as having exactly the same meaning, i.e. in general,

$$\frac{dy}{dx} \equiv \lim_{\delta x \to 0} \frac{\delta y}{\delta x}$$

This is called the *derivative of y with respect to x*; it can also be described as the result of *differentiating y* with respect to x. Quantities that are arbitrarily close to zero while remaining distinguishable from zero are called *infinitesimal*.

To summarize this section, therefore, we have examined the function

$$y = 2 + 3x + x^2$$

and found its derivative with respect to x to be

$$\frac{dy}{dx} = 3 + 2x$$

....

Example 4.1 Differentiation from first principles
Differentiate the equation $y = x/(1+x)$ with respect to x from first principles.

For a small change δx in x that produces a small change δy in y,

$$y + \delta y = \frac{x + \delta x}{1 + x + \delta x}$$

The change in y is then $y + \delta y - y$, i.e.

$$\delta y = \frac{x + \delta x}{1 + x + \delta x} - \frac{x}{1 + x}$$

$$= \frac{(x + \delta x)(1 + x) - x(1 + x + \delta x)}{(1 + x + \delta x)(1 + x)}$$

$$= \frac{\delta x}{(1 + x + \delta x)(1 + x)}$$

and dividing both sides by δx we have

$$\frac{\delta y}{\delta x} = \frac{1}{(1 + x + \delta x)(1 + x)}$$

When δx is vanishingly small $(1 + x + \delta x)$ is indistinguishable from $(1 + x)$, so the derivative is

$$\frac{dy}{dx} = \lim_{\delta x \to 0} \frac{\delta y}{\delta x} = \frac{1}{(1 + x)^2}$$

....

4.3 **Rapid differentiation**

The method we used in the previous section for differentiating $y = 2 + 3x + x^2$ is known as *differentiating from first principles*. It is quite general, i.e. the same approach can be used for differentiating any function. That is to say, one can differentiate any function whatsoever by using it to get expressions for y and $y + \delta y$ at x and $x + \delta x$, respectively, subtracting one from the other to get an expression for δy in terms of x and δx, dividing δy by δx and examining, by inspection, the form of the expression approached when δx is made infinitesimal. We shall use this method again in this chapter to see how to obtain formulae for differentiating products, ratios, and other functions. However, for everyday use it is rather long-winded, and in fact there are rules for the commonest cases that are simple enough to make it unnecessary to go through the whole process of differentiation from first principles.

In particular, there is a simple rule for differentiating any power of x with respect to x. In general, if

$$y = Ax^i$$

where A and i are constants, then

$$\frac{dy}{dx} = Aix^{i-1}$$

Putting this into words, any constant multiplier (A in this case) remains unchanged, the index (i) is decreased by 1 (giving $i-1$) and the result is multiplied by the old index (i). Applying this rule to various examples, with A representing a constant in each case, we have

$$\frac{dA}{dx} = 0$$

$$\frac{d}{dx}(Ax) = A$$

$$\frac{d}{dx}(Ax^2) = 2Ax$$

$$\frac{d}{dx}(Ax^3) = 3Ax^2$$

$$\frac{d}{dx}\left(\frac{A}{x}\right) = \frac{d}{dx}(Ax^{-1}) = -Ax^{-2} = -\frac{A}{x^2}$$

$$\frac{d}{dx}(Ax^{1/2}) = \frac{1}{2}Ax^{-1/2} = \frac{A}{2x^{1/2}}$$

and so on. Notice that it is *not necessary* (and pointless) to try to remember all these separately, because they all follow quite mechanically from the general expression given first.

There are, however, two special cases, not of the above form, that occur so often that they need to be committed to memory. These are the rules for differentiating logarithms and exponentials:

$$\frac{d}{dx}(A\ln x) = \frac{A}{x}$$

$$\frac{d}{dx}(Ae^x) = Ae^x$$

It is common to memorize them in the special cases with $A=1$, and this is acceptable as long as one remembers how to go to the more general case where $A \neq 1$:

$$\frac{\mathrm{d}}{\mathrm{d}x}(\ln x) = \frac{1}{x}$$

$$\frac{\mathrm{d}}{\mathrm{d}x}(\mathrm{e}^x) = \mathrm{e}^x$$

When supplemented with the rules for differentiating combinations of terms that are set in the next three sections, the general expression for powers of x and the two special cases just noted allow differentiation of almost all the expressions encountered in elementary biochemistry.

4.4 Derivatives of sums and products

We already considered the derivative of a sum when we differentiated $y = 2 + 3x + x^2$ from first principles in Section 4.2, and we found the result to be as follows:

$$\frac{\mathrm{d}y}{\mathrm{d}x} = 3 + 2x$$

If we write this as

$$\frac{\mathrm{d}y}{\mathrm{d}x} = 0 + 3 + 2x$$

we can see by inspection that each term in the answer is the derivative of the corresponding term in the initial equation, i.e.

$$\frac{\mathrm{d}}{\mathrm{d}x}(2) = 0, \qquad \frac{\mathrm{d}}{\mathrm{d}x}(3x) = 3, \qquad \frac{\mathrm{d}}{\mathrm{d}x}(x^2) = 2x$$

and so

$$\frac{\mathrm{d}}{\mathrm{d}x}(2 + 3x + x^2) = \frac{\mathrm{d}}{\mathrm{d}x}(2) + \frac{\mathrm{d}}{\mathrm{d}x}(3x) + \frac{\mathrm{d}}{\mathrm{d}x}(x^2)$$

In words, *the derivative of the sum is the sum of the derivatives*. The question then arises whether this is a chance result with a particular example or is it always true. In fact it is always true, as we can see by going back to first principles.

Suppose we wish to differentiate a sum, such as $y = u + v$, where u and v are both functions of x that can readily be differentiated with respect to x: by the arguments we have used before, a small change δx in x brings about changes in u and v as follows:

$$\delta u \approx \frac{\mathrm{d}u}{\mathrm{d}x}\,\delta x, \qquad \delta v \approx \frac{\mathrm{d}v}{\mathrm{d}x}\,\delta x$$

As $y = u + v$, the change in y must be the sum of the changes in u and v:

$$\delta y = \delta u + \delta v$$

and dividing by δx we have

$$\frac{\delta y}{\delta x} = \frac{\delta u}{\delta x} + \frac{\delta v}{\delta x} \approx \frac{du}{dx} + \frac{dv}{dx}$$

In the limit, the approximation becomes exact, i.e.

$$\frac{dy}{dx} = \lim_{\delta x \to 0} \frac{\delta y}{\delta x} = \frac{du}{dx} + \frac{dv}{dx}$$

This result can be extended to the sums with any number of terms, and in general, therefore, the derivative of a sum is equal to the sum of the derivatives.

The derivative of a difference follows directly:

$$\frac{d}{dx}(u - v) = \frac{du}{dx} - \frac{dv}{dx}$$

as a difference is simply a sum in which the second term is negative.

Suppose now that $y = uv$ is the product of two functions of x. Then

$$y + \delta y = (u + \delta u)(v + \delta v) = uv + v\,\delta u + u\,\delta v + \delta u\,\delta v$$

Remembering that $y = uv$, we can subtract y from the left-hand side and uv from the right-hand side to get an expression for δy:

$$\delta y = v\,\delta u + u\,\delta v + \delta u\,\delta v$$

Now it should be evident that as we make δx smaller all of the terms in this equation become smaller also, but as the last term is the product of two small numbers it approaches zero much more rapidly than the others. So we can write, approximately:

$$\delta y \approx v\,\delta u + u\,\delta v$$

Dividing all terms by δx, therefore, we have:

$$\frac{\delta y}{\delta x} \approx v\frac{\delta u}{\delta x} + u\frac{\delta v}{\delta x}$$

In the limit, this becomes exact, and we can replace all the ratios by derivatives:

$$\frac{dy}{dx} = \lim_{\delta x \to 0} \frac{\delta y}{\delta x} = v\frac{du}{dx} + u\frac{dv}{dx}$$

This result is often memorized in the form $D(uv) = u\,Dv + v\,Du$, in which D is an abbreviation for the differentiation operator $\frac{d}{dx}$.

The derivation is illustrated graphically in Fig. 4.4: the change in uv is represented by the sum of the areas in the three rectangles labelled $u\,\delta v$, $v\,\delta u$, and $\delta u\,\delta v$,

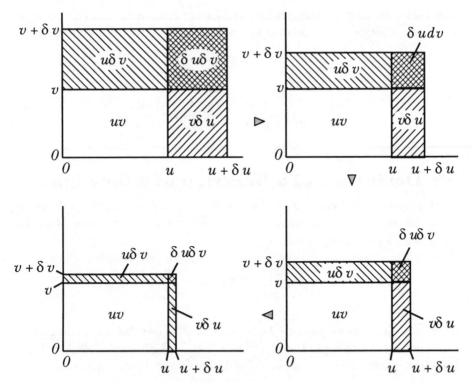

Fig. 4.4 Explanation of the formula for the derivative of a product uv. When the increments δu and δv are large, their product $\delta u\,\delta v$ is comparable in size with $u\,\delta v + v\,\delta u$ (*top-left*); but as they become smaller (*top-right* and *bottom-right*), $\delta u\,\delta v$ becomes smaller faster than $u\,\delta v + v\,\delta u$, and when they are very small, $\delta u\,\delta v$ is negligible compared with $u\,\delta v + v\,\delta u$ (*bottom-left*).

and if we follow the figure in a clockwise direction starting from the top-left we see that the whole diagram is drawn four times for progressively smaller values of δx. When δx is quite large (top-left), all three rectangles are correspondingly large. It is noticeable that the rectangle labelled $\delta u\,\delta v$ is not much smaller than the other two; but as δx is made smaller, it decreases in size much more rapidly than they do, so that by the time δx is very small (bottom-left) it has almost disappeared even though the others have not. In other words, its area approaches zero much more rapidly than they do, and $\delta u\,\delta v$ becomes negligible when $u\,\delta v$ and $v\,\delta u$ are not negligible.

Although the expression above for the derivative of a product is usable as it stands, it has a property that is not very convenient if we want to generalize it to more complex cases: the terms on the right-hand side are not 'homogeneous', which means that each of them contains three kinds of variables, x and two others, u and v. We can make all the terms in the equation homogeneous if we divide the left-hand side by y and the right-hand side by uv (remember that uv is the same as y):

$$\frac{1}{y}\frac{dy}{dx} = \frac{1}{u}\frac{du}{dx} + \frac{1}{v}\frac{dv}{dx}$$

Now each term contains only one kind of variable apart from x, and it is easier to see what will happen if the product has more than two terms. For example, if

$$y = uvw$$

then

$$\frac{1}{y}\frac{dy}{dx} = \frac{1}{u}\frac{du}{dx} + \frac{1}{v}\frac{dv}{dx} + \frac{1}{w}\frac{dw}{dx}$$

4.5 Derivative of a 'function of a function'

The bell-shaped curves often found for the pH-dependence of protein and enzyme properties (such as catalytic activity) can be represented by an equation of the following sort, known as a *Michaelis function*:

$$y = \frac{A}{1 + \dfrac{B}{x} + Cx} = A(1 + Bx^{-1} + Cx)^{-1}$$

where y is the protein property being studied, x is (usually) the hydrogen-ion concentration, i.e. not pH but 10^{-pH} (see Section 3.12), and A, B and, C are constants. A new variable u can be defined to represent the denominator:

$$u = 1 + Bx^{-1} + Cx$$

and this can be used to write the expression as a whole as follows:

$$y = Au^{-1}$$

It is now easy to differentiate u with respect to x:

$$\frac{du}{dx} = -Bx^{-2} + C$$

and equally easy to differentiate y with respect to u:

$$\frac{dy}{du} = -Au^{-2}$$

but not so obvious how to differentiate y with respect to x: how can we combine the two results to get this?

This is an example of how to differentiate a *function of a function*: y is a function of u, which is itself a function of x. I shall now examine this as a general problem, and later we shall return to see how the general solution can be applied to the specific case of the Michaelis function.

By the same logic as we have used already in this chapter, a small increment δy in y can be expressed approximately in terms of the corresponding increment δu in u by using the derivative with respect to x. Then δu can be expressed similarly in terms of δx:

$$\delta y \approx \frac{dy}{du} \delta u, \qquad \delta u \approx \frac{du}{dx} \delta x$$

and so, substituting for δu,

$$\delta y \approx \frac{dy}{du} \frac{du}{dx} \delta x$$

and dividing both sides by δx:

$$\frac{\delta y}{\delta x} \approx \frac{dy}{du} \frac{du}{dx}$$

In the limit, this becomes exact:

$$\frac{dy}{dx} = \lim_{\delta x \to 0} \frac{\delta y}{\delta x} \approx \frac{dy}{du} \frac{du}{dx}$$

In general, if f_1 is a function of f_2, f_2 is a function of f_3, etc., to f_n, which is a function of x, then

$$\frac{df_1}{dx} = \frac{df_1}{df_2} \frac{df_2}{df_3} \frac{df_3}{df_4} \cdots \frac{df_n}{dx}$$

This is known as the *chain rule*. It may seem obvious, as it can be 'derived' by treating all of the derivatives as fractions and cancelling the like elements. Although such a 'derivation' would not satisfy a serious mathematician, the differences between derivatives and fractions rarely create difficulties in biochemistry. Consequently, it does little harm if the biochemist uses non-rigorous 'derivations' as an aid to remembering important general relationships.

Returning to the Michaelis function, application of the chain rule shows the derivative to be

$$\frac{dy}{dx} = \frac{dy}{du} \frac{du}{dx} = -Au^{-2}(-Bx^{-2} + C)$$

and the process can be completed by substituting back the expression for u in terms of x:

$$\frac{dy}{dx} = \frac{ABx^{-2} - AC}{(1 + Bx^{-1} + Cx)^2}$$

4.6 **Derivative of a ratio**

A ratio can be regarded as a product in which the second term is raised to the power -1. So $y = u/v$ is the same as $y = uv^{-1}$, and

$$\frac{dy}{dx} = u \frac{d}{dx}(v^{-1}) + v^{-1} \frac{du}{dx}$$

in which v^{-1} is a function of a function (it is a function of v, which is a function of x), so it can be differentiated as in the previous section:

$$\frac{d}{dx}(v^{-1}) = \frac{d}{dv}(v^{-1})\frac{dv}{dx} = -v^{-2}\frac{dv}{dx}$$

and substituting this into the first expression we have the derivative of y with respect to x:

$$\frac{dy}{dx} = -uv^{-2}\frac{dv}{dx} + v^{-1}\frac{du}{dx} = \frac{v\dfrac{du}{dx} - u\dfrac{dv}{dx}}{v^2}$$

Using the same sort of notation as we introduced earlier for the derivative of a product $[D(uv) = u\,Dv + v\,Du]$, we can think of this result as $D(u/v) = (v\,Du - u\,Dv)/v^2$.

....

Example 4.2 Differentiation of a ratio

Differentiate the equation $y = x/(1+x)$ with respect to x by using the formula for the derivative of a ratio.

Define $u = x$, $v = (1+x)$, then

$$\frac{du}{dx} = 1; \qquad \frac{dv}{dx} = 1$$

and so

$$\frac{dy}{dx} = \frac{v\dfrac{du}{dx} - u\dfrac{dv}{dx}}{v^2} = \frac{v-u}{v^2} = \frac{1+x-x}{(1+x)^2} = \frac{1}{(1+x)^2}$$

This is, of course, the same as the result obtained in Example 4.1 by differentiating from first principles.

....

4.7 Higher derivatives

After a function has been differentiated once, there is no reason why the result should not be differentiated a second time by application of exactly the same rules. For example,

$$y = 7 + 3x - 2x^2 + x^3$$

$$\frac{dy}{dx} = 3 - 4x + 3x^2$$

$$\frac{d}{dx}\left(\frac{dy}{dx}\right) = -4 + 6x$$

$$\frac{d}{dx}\left[\frac{d}{dx}\left(\frac{dy}{dx}\right)\right]=6$$

$$\frac{d}{dx}\left\{\frac{d}{dx}\left[\frac{d}{dx}\left(\frac{dy}{dx}\right)\right]\right\}=0$$

Although the expression on the second line is often called just the derivative, it is more precise to call it the *first derivative*, and this terminology can be extended in an obvious way to the next line, which shows the *second derivative*, and similarly for the third and fourth derivatives shown on the next lines. As we can see in the example, applying the first-derivative notation to the higher derivatives rapidly generates rather cumbersome symbols, and in practice we define special symbols for the higher derivatives:

$$\frac{d^2y}{dx^2}\equiv\frac{d}{dx}\left(\frac{dy}{dx}\right)$$

$$\frac{d^3y}{dx^3}\equiv\frac{d}{dx}\left(\frac{d^2y}{dx^2}\right)$$

etc. (These symbols are more logical than they may seem at first sight if we accept that the exponents in the denominator act on the whole expression dx and not just on x: remember that the d is an operator, not a variable.)

What do these higher derivatives mean? Just as the first derivative $\frac{dy}{dx}$ expresses how fast y changes with x, the second derivative expresses how fast $\frac{dy}{dx}$ changes with x. Alternatively, if $\frac{dy}{dx}$ expresses the slope of a plot of y against x, $\frac{d^2y}{dx^2}$ expresses the rate of change of slope, or in other words the *curvature*. If a line is straight its curvature is zero (that is what 'straight' means) and its slope is constant, so the second derivative is zero. For a general straight line, therefore,

$$y=A+Bx$$

$$\frac{dy}{dx}=B$$

$$\frac{d^2y}{dx^2}=0$$

4.8 Notation

The notation used hitherto in this chapter is based on that of the great German mathematician Leibniz. It is the most generally useful and the most widely used, but there are circumstances in which it becomes cumbersome, and one needs some familiarity with three other systems that are sometimes used:

(1) Mathematical tables are used to tabulate standard expressions, so that if one has forgotten the form of a particular result one can look it up in the appropriate table. In tables of derivatives it is usually obvious what variable one is differentiating with respect to, and compactness is achieved by writing $\frac{dy}{dx}$ just as D. This notation has already been mentioned in this chapter for memorizing the standard results for the derivatives of products and ratios:

$$D(uv) = u\,Dv + v\,Du$$

$$D\left(\frac{u}{v}\right) = \frac{v\,Du - u\,Dv}{v^2}$$

(2) Another system for achieving compactness when it is obvious what one is differentiating with respect to is to write the first derivative of y with respect to x as y', and the second derivative as y''. This is especially convenient when one does not want to define a particular variable (y in this example) as the function, for example, one can write the first derivative of $f(x)$ as $f'(x)$, and the second as $f''(x)$.

(3) Calculus was developed independently by Newton in England and Leibniz in Germany, and so it is hardly surprising that Newton developed his own *dot notation*, which is not the same as Leibniz's notation. For most purposes, Newton's notation is much less convenient, in part because it does not allow for the possibility that one may wish to differentiate with respect to more than one variable. Newton was mainly concerned with functions of *time*, or *t*, and for his purposes it was possible to assume that all differentiation was with respect to *t*. When dealing with time (e.g. in enzyme kinetics), it is still occasionally convenient to write $\dot{x}$ and $\ddot{x}$ to mean the first and second derivatives, respectively, of x with respect to t. Even in kinetics, however, this notation is much less common than that of Leibniz.

The various alternative systems are listed in Table 4.3. As indicated in the table, one can continue differentiating beyond the second derivative, but the biochemist rarely wants to do this; so, I have not discussed it in the text. It will be

Table 4.3 Notations for writing derivatives.

Function	y	y	y	$f(x)$	$x = f(t)$
1st derivative	$\dfrac{dy}{dx}$	Dy	y'	$f'(x)$	$\dot{x}$
2nd derivative	$\dfrac{d^2y}{dx^2}$	D^2y	y''	$f''(x)$	$\ddot{x}$
nth derivative	$\dfrac{d^n x}{dx^n}$	$D^n y$	$y^{(n)}$	$f^{(n)}(x)$	—

noticed that although Leibniz's notation is the least compact, it is also the most explicit. For this reason, it is nearly always wise for the non-mathematician to prefer it: a small saving in space is never justified if the result is incomprehension.

4.9 Maxima

The equation

$$v = \frac{Va}{K_m + a + \dfrac{a^2}{K_{si}}}$$

represents a simple kind of *substrate inhibition* in an enzyme-catalysed reaction. The variables are the rate, v, and the concentration of substrate, a, and V, K_m, and K_{si} are constants. Unlike the examples earlier in this chapter, this equation is not written with the typical symbols of elementary mathematics, x, y, A, B, etc., but with the sort of symbols most likely to be encountered in biochemistry. This difference should cause no problems: one does not require a particular set of symbols to carry out mathematical operations, and it is unwise to associate particular relationships too rigidly with particular symbols.

Figure 4.5 shows the form of a plot of v against a that the equation defines for particular values of the constants. A striking feature of the curve is that, instead of increasing indefinitely with a (as in simpler kinds of kinetic behaviour), v increases to a *maximum* and then decreases when a is increased further. A major use of the differential calculus is that it enables one to calculate where a maximum will occur. From inspection of the graph it can be seen that the slope is zero at the maximum. This is true in general, i.e. at any maximum the slope is zero

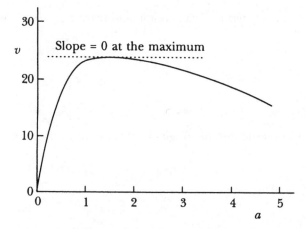

Fig. 4.5 Plot of rate v against substrate concentration a for a system showing substrate inhibition. The plot illustrates the property that at the maximum the slope is zero. However, the existence of a zero slope is not a definition of a maximum because it is also a property of a minimum (see Fig. 4.6).

(although the converse is not true, because a point of zero slope may also represent a minimum or a point of inflection). Accordingly, we can find the maximum in Fig. 4.5 by differentiating v with respect to a and finding a value of a at which the result is zero. This can be done by applying the rule for differentiating a ratio:

$$\frac{dv}{da} = \frac{\left(K_m + a + \dfrac{a^2}{K_{si}}\right)V - Va\left(1 + \dfrac{2a}{K_{si}}\right)}{\left(K_m + a + \dfrac{a^2}{K_{si}}\right)^2}$$

$$= \frac{VK_m + Va + \dfrac{Va^2}{K_{si}} - Va - \dfrac{2Va^2}{K_{si}}}{\left(K_m + a + \dfrac{a^2}{K_{si}}\right)^2}$$

and collecting like terms in the numerator gives

$$\frac{dv}{da} = \frac{VK_m - \dfrac{Va^2}{K_{si}}}{\left(K_m + a + \dfrac{a^2}{K_{si}}\right)^2}$$

In principle, an expression such as this can be zero either because the numerator is zero or because the denominator is infinite. Although the denominator of this particular expression does increase steeply and without limit as a increases, it remains finite at all finite a. So the only way the expression can be zero at finite a is for the numerator to be zero, i.e.

$$VK_m - \frac{Va^2}{K_{si}} = 0$$

As V is a factor of every term we can omit it, and rearrange what is left to provide an expression for a^2,

$$a^2 = K_m K_{si}$$

and hence

$$a = (K_m K_{si})^{1/2}$$

Substitution of this result into the original equation shows the maximum value of v to be

$$v = \frac{V(K_m K_{si})^{1/2}}{K_m + (K_m K_{si})^{1/2} + \dfrac{K_m K_{si}}{K_{si}}}$$

$$= \frac{V(K_m K_{si})^{1/2}}{(K_m K_{si})^{1/2} + 2K_m} = \frac{V}{1 + 2(K_m/K_{si})^{1/2}}$$

The Michaelis function that we considered earlier in this chapter has exactly the same form as the equation for substrate inhibition, because if

$$y = \frac{A}{1 + \dfrac{B}{x} + Cx}$$

then multiplication of both numerator and denominator by x gives

$$y = \frac{Ax}{B + x + Cx^2}$$

which is exactly the same equation with different symbols, as we may readily see by replacing y with v, x with a, A with V, B with K_m, and C with $1/K_{si}$. It follows that the maximum of the Michaelis function must occur when

$$x = \left(\frac{B}{C}\right)^{1/2}, \qquad y = \frac{A}{1 + 2(BC)^{1/2}}$$

In this case, however, we are more likely to want to plot y against $\log x$ than against x (because the Michaelis function is a dependence on hydrogen-ion concentration, which is usually expressed in logarithmic form). Thus, question arises, can we assume that the same maximum occurs in a plot of y against $\log x$? Put differently, can we assume that

$$\frac{dy}{d\log x} = 0 \quad \text{when} \quad \frac{dy}{dx} = 0?$$

The answer comes from considering the relationship between the two derivatives, which follows from the chain rule:

$$\frac{dy}{d\log x} = \frac{dy}{dx} \frac{dx}{d\ln x} \frac{d\ln x}{d\log x}$$

in which

$$\frac{dx}{d\ln x} = x \quad \text{and} \quad \frac{d\ln x}{d\log x} \approx 2.303$$

because

$$\frac{d\ln x}{dx} = \frac{1}{x} \quad \text{and} \quad \ln x \approx 2.303 \log x$$

and, substituting, we find:

$$\frac{dy}{d\log x} \approx 2.303x \frac{dy}{dx}$$

Thus, provided x is finite, a zero value of $\dfrac{dy}{d\log x}$ must occur at the same value of x as a zero value of $\dfrac{dy}{dx}$, and so the maximum in the plot of y against $\log x$ occurs at the same values of x and y as in the plot of y against x.

4.10 **Minima**

Exactly the same criteria can be used to find a *minimum* of a function, because this also corresponds to a zero value of the first derivative (Fig. 4.6). Clearly, therefore, the existence of such a zero value does not guarantee that a maximum has been found. It shows only that one has found a *stationary point*, a more general term that embraces both maxima and minima (and some other types of point that we can ignore here). In most elementary situations in biochemistry, this is not a problem because it is usually obvious whether one is dealing with a maximum or a minimum. In the more general case, there may be ambiguity, but this can be resolved by examining the sign of slope on either side of the stationary point. At a maximum, the slope is *decreasing* through zero, whereas at a minimum it is *increasing* through zero, as illustrated in Fig. 4.6.

We can apply these criteria to the expression for $\frac{dv}{da}$ in the case of substrate inhibition: it is clear that the numerator, $VK_m - Va^2/K_{si}$, can only decrease if a is increased (for any positive value of a), and so it must decrease through zero as a increases through the value at the stationary point. This is, therefore, a maximum, as assumed previously, not a minimum. It is not necessary to consider the denominator of the expression for $\frac{dv}{da}$ because this contains positive terms only (for positive a) and must therefore be positive.

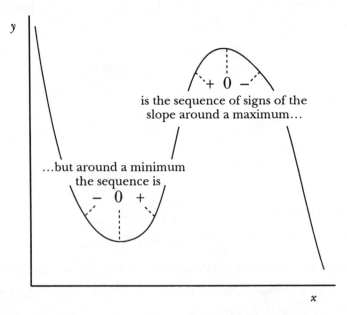

Fig. 4.6 Although the slope is zero both at a minimum and at a maximum, the progression of signs in the vicinity is different in the two cases: $+0-$ as the abscissa variable increases for a maximum, but $-0+$ for a minimum.

Example 4.3 Finding a stationary point
Find a value of x at which the function $y = 4 - 3x + 2x^2$ has a stationary point, and identify whether it is a maximum or a minimum.

Differentiating gives

$$\frac{dy}{dx} = -3 + 4x$$

which is zero when $4x = 3$, i.e. $x = \frac{4}{3}$. When x is a little smaller than this value, the derivative is negative, and when x is a little larger than this value the derivative is positive. The slope thus changes from negative to positive as x increases through the stationary point, which is therefore a minimum.

Alternatively, we can differentiate a second time:

$$\frac{d^2 y}{dx^2} = 4$$

and obtain a second derivative that is positive for all values of x, implying that the slope can only increase and so any stationary point must be a minimum.

4.11 **A note on terminology**

Mathematicians have a greater need for precise and unambiguous meanings for the terms they use than we have in everyday life. As a result, terms in mathematics often have exact meanings that do not precisely agree with the meanings of the same words when used in everyday language. An example is the term 'maximum', which refers in mathematics, as we have seen, to a point at which the first derivative of a function is zero and is decreasing as the abscissa variable increases. For a complex function, there may be many such points, and in such a case we would refer to them as 'maxima' and to one of them as 'a maximum', not as 'the maximum'. Similar considerations apply to minima. Even if the curve has a unique maximum, it can continue to values greater than this maximum if there is also a minimum, and likewise it can continue to values smaller than the minimum: both properties are illustrated in Fig. 4.6, as the values of y at the lowest x values are greater than the maximum, and the values of y at the highest x values are less than the minimum. Moreover, there is nothing in the definitions that requires a maximum to occur at a larger value of the function than a minimum; indeed, quite simple functions exist with a unique maximum at a lower value than the unique minimum, such as

$$y = x + \frac{1}{x}$$

(If you find the description of the curve unbelievable, you should make a plot of the equation for x between -2 and -0.1 and between 0.1 and 2.)

Several of the statements in this section would be absurd in everyday language, in which the maximum of a measurement is the largest value it can have and the minimum is the smallest value it can have, with 'it can have' usually taken to include the qualification 'under realistic circumstances'.

Because biochemistry was not developed largely by mathematicians, mathematical terms have come to be used in biochemistry with looser meanings than would be acceptable in mathematics. For example, the quantity V in the Michaelis–Menten equation:

$$v = \frac{Va}{K_m + a}$$

is often called the *maximum velocity*, because v cannot exceed it. However, v cannot attain the value V at a finite value of a and thus V is not a maximum in the mathematical sense but a *limit*. A better name for V, therefore, is the *limiting rate*, a term gaining currency among biochemists, though still far from universal. When a true maximum does occur in a plot of v against a, as in the example of substrate inhibition discussed in Section 4.9, the value of v at the maximum should not be called the 'maximum velocity', because this would cause confusion with the usual meaning given to this term.

4.12 **Points of inflection**

There is an alternative to the method described in Section 4.10 for deciding whether a stationary point is a maximum or a minimum; instead one can examine the sign of the second derivative at the stationary point. This is because the second derivative expresses directly how the first derivative is changing with the abscissa variable. At a maximum, the first derivative is decreasing and so the second derivative is normally negative; conversely, at a minimum, it is normally positive. Although I could illustrate this with the same example as in Section 4.10, we would then get a rather complicated expression on differentiating twice; it is easier to consider a function such as the following:

$$y = 2 + 9x - 6x^2 + x^3$$

for which

$$\frac{dy}{dx} = 9 - 12x + 3x^2$$

$$\frac{d^2y}{dx^2} = -12 + 6x$$

The first derivative is zero when $x=1$ and $y=6$, and also when $x=3$ and $y=2$. As 6 is larger than 2, it may seem obvious that the first solution is a maximum and the second a minimum. However, although this conclusion happens to be correct in this instance it does not follow from the argument, which is unsound (recall the example in the previous section with a unique minimum at a higher value than the unique maximum). It is safer, therefore, to examine the value of the second derivative: this is -6 at $x=1$ and $+6$ at $x=3$, and shows that the point $(1,6)$ is indeed a maximum and that the point $(3,2)$ is a minimum.

Occasionally, the first and second derivatives are both zero at the same value of x, but this is unusual in the sort of equations that occur in biochemistry and thus rarely interferes with the method just described for distinguishing between maxima and minima. It is, however, important to consider zero values of the second derivative at non-zero values of the first derivative. These occur at so-called *points of inflection*, which are points at which the slope of the plot is a maximum or a minimum (Fig. 4.7).

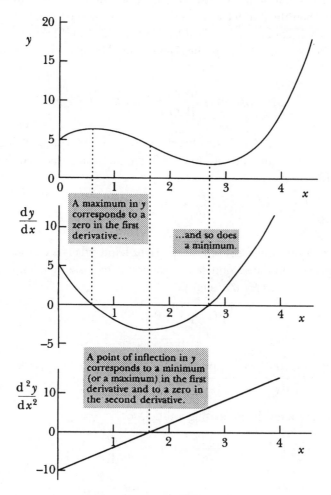

Fig. 4.7 Point of inflection. At a maximum or a minimum, the first derivative is zero; at a point of inflection the first derivative is a maximum or a minimum and the second derivative is zero.

Points of inflection are important in biochemistry because they define conditions in which a response (e.g. the rate of a reaction) is most (or least) sensitive to an influence (e.g. the concentration of a metabolite). Consider, for example, the ability of a buffer to resist the changes in pH that might be brought about by addition of alkali. The equation that defines this (to a first approximation) is the *Henderson–Hasselbalch equation*:

$$pH = pK_a + \log \frac{[\text{salt}]}{[\text{acid}]}$$

in which pK_a is the negative logarithm of the acid dissociation constant K_a, [acid] is the concentration of weak acid in the mixture and [salt] is the concentration of a salt of the same acid. For example, for a buffer made up by adding $x \, \text{mol} \, L^{-1}$ of NaOH to a solution of acetic acid, HOAc, at an initial concentration $A \, \text{mol} \, L^{-1}$ (with $A > x$), then the NaOH will react with a stoichiometric amount of HOAc, producing a salt concentration [NaOAc] of $x \, \text{mol} \, L^{-1}$ and leaving an acid concentration of $A - x \, \text{mol} \, L^{-1}$, so the Henderson–Hasselbalch equation for this system is

$$pH = pK_a + \log \frac{[\text{NaOAc}]}{[\text{HOAc}]} = pK_a + \log \frac{x}{A-x}$$

It will be easier to differentiate this if we replace the common logarithm by the natural logarithm and, as $\ln x = 2.303 \log x$, it follows that $\log x = (\ln x)/2.303 = 0.4343 \ln x$:

$$pH = pK_a + 0.4343 \ln \frac{x}{A-x}$$

Differentiating this, the first term on the right-hand side is a constant and therefore disappears, and the second can be differentiated as a function of a function. If we write $x/(A-x)$ as $f = u/v$, then

$$pH = pK_a + 0.4343 \ln f$$

and

$$\frac{df}{dx} = \frac{(A-x)(1) - x(-1)}{(A-x)^2} = \frac{A-x+x}{(A-x)^2} = \frac{A}{(A-x)^2}$$

(from the formula for differentiating a ratio), so

$$\frac{dpH}{dx} = 0.4343 \, \frac{1}{f} \, \frac{df}{dx}$$

$$= 0.4343 \times \frac{A-x}{x} \times \frac{A}{(A-x)^2}$$

$$= \frac{0.4343A}{x(A-x)}$$

This first derivative is a measure of the sensitivity of the pH to addition of base: if it is small, it means that adding a trace of alkali will not change the pH very much. Clearly, this is what we expect of an effective buffer; so, to determine the pH at which the buffer is most effective, we need to know the pH at which the derivative is a minimum (or at which the reciprocal of this derivative, known as the *buffer capacity*, is a maximum).

Finding the value of x that makes the derivative a minimum requires a second differentiation. This is again most easily done by recognizing that we are dealing with a function of a function: if we define $g = x(A-x) = Ax - x^2$, then

$$\frac{dpH}{dx} = 0.4343\,Ag^{-1}, \quad g = x(A-x)$$

$$\frac{dg}{dx} = A - 2x$$

$$\frac{d^2pH}{dx^2} = \left(\frac{d}{dg}\frac{dpH}{dx}\right)\frac{dg}{dx} = -0.4343Ag^{-2}(A-2x)$$

$$\frac{d^2pH}{dx^2} = \left(\frac{d}{dg}\frac{dpH}{dx}\right)\frac{dg}{dx} = -0.4343Ag^{-2}(A-2x)$$

and we can conclude by replacing g by its definition, $x(A-x)$:

$$\frac{d^2pH}{dx^2} = \frac{-0.4343A(A-2x)}{(Ax - x^2)^2}$$

By inspection, it is clear that this second derivative is zero when $x = A/2$, i.e. when enough alkali has been added to neutralize exactly half of the original acid. Substituting back into the Henderson–Hasselbalch equation, we find, remembering that $\ln(1) = 0$, that

$$pH = pK_a + 0.4343 \ln\frac{A/2}{A - A/2} \quad pK_a + 0.4343 = \ln\left(\frac{2A}{2A}\right) = pK_a$$

Thus, a buffer is most effective at the pH equal to the pK_a of the acid involved in the titration.

4.13 Sketching curves

One of the most important uses of differential calculus is as an aid in sketching the curves corresponding to unfamiliar functions. For an unknown function, if we know where it crosses the axis (at the values of x that give zero values of the function), and where it has minima, maxima and points of inflection then we ought to be able to deduce a clear idea of what the curve looks like.

Suppose, for example that one had found that the concentration q of product of a reaction at time t could be expressed by an equation of the following form, in which all symbols apart from q and t are positive constants:

$$q = \frac{k_2 k_3 e_0 t}{k_2 + k_3} - \frac{k_2 k_3 e_0 \{1 - \exp[-(k_2 + k_3)t]\}}{(k_2 + k_3)^2}$$

It is by no means obvious at first glance what sort of curve this equation defines, but one can deduce a good deal simply by examining two extreme values of t. First, if $t = 0$ then

$$q = -\frac{k_2 k_3 e_0 [1 - \exp(0)]}{(k_2 + k_3)^2}$$

However, we know from Chapter 3 that any number raised to the power 0 has a value of 1, so $\exp(0) = 1$ and $1 - \exp(0) = 0$. Thus,

$$q = 0$$

which means that the curve passes through the origin.

Second, as t is positive the product $-(k_2 + k_3)t$ is always negative and increases in magnitude in proportion to t. Thus, $\exp[-(k_2 + k_3)t]$ is the number e raised to a negative power, or the reciprocal of e raised to a positive power. Now it is not easy to do mental arithmetic with $e = 2.718...$, but we can get an idea of how it will change by looking at successive powers of 2 instead: $2, 4, 8, 16, 32 \ldots$. These increase very steeply, so their reciprocals decrease correspondingly steeply (while always remaining positive). It is clear therefore that as t increases $\exp[-(k_2 + k_3)t]$ will get closer and closer to zero, and when it is small enough we can ignore it, and the equation will simplify to

$$q = \frac{k_2 k_3 e_0 t}{k_2 + k_3} - \frac{k_2 k_3 e_0}{(k_2 + k_3)^2}$$

The first term on the right-hand side is just a positive constant multiplied by t, and the second is a negative constant. So this is the expression of a straight line with a positive slope and a negative intercept on the ordinate. It follows that the curve defined by the complete equation must approximate to such a straight line at large values of t.

Although this information is useful, it tells us little about what might well prove to be the most interesting part of the time course, the period before the straight-line approximation becomes valid. Light can be shed on this period by differentiating twice to reveal the behaviour of the slope:

$$\frac{dq}{dt} = \frac{k_2 k_3 e_0}{k_2 + k_3} - \frac{k_2 k_3 e_0 \exp[-(k_2 + k_3)t]}{k_2 + k_3}$$

$$= \frac{k_2 k_3 e_0 \exp\{1 - \exp[-(k_2 + k_3)t]\}}{k_2 + k_3}$$

$$\frac{d^2 q}{dt^2} = k_2 k_3 e_0 \exp[-(k_2 + k_3)t]$$

As $\exp[-(k_2 + k_3)t]$ must have a value between 1 (at $t=0$) and 0 (for t approaching infinity) for any positive value of t, it follows that the first derivative cannot be negative and is zero only at the origin. Thus, the plot of q against t cannot contain a maximum or a minimum except at the origin. Moreover, the second derivative is positive at all values of t, albeit with a value that becomes negligible at large values of t. When all of this information is combined, it becomes clear that the form of the curve must be as shown in Fig. 4.8.

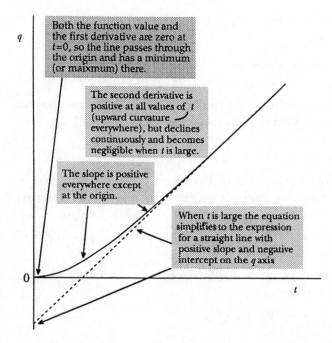

Both the function value and the first derivative are zero at $t=0$, so the line passes through the origin and has a minimum (or maixmum) there.

The second derivative is positive at all values of t (upward curvature everywhere), but declines continuously and becomes negligible when t is large.

The slope is positive everywhere except at the origin.

When t is large the equation simplifies to the expression for a straight line with positive slope and negative intercept on the q axis

Fig. 4.8 Sketching a curve. By considering the values of a function and its derivatives at various points, where they are easy to calculate, one can deduce the shape that a curve must have (see the text for details).

4.14 **Problems**

4.1 Differentiate each of the following expressions with respect to x:

(a) $y = x^3$,　　　　　　(b) $y = x^{1/2}$

(c) $y = 5e^x$,　　　　　　(d) $y = e^{5x}$

(e) $y = e^3$,　　　　　　(f) $y = 3 \ln x$

(g) $y = x^{1/2} + x^{-1/2}$,　(h) $y = \ln x - 2x^2$

(i) $y = x^2 e^x$,　　　　　(j) $y = \ln(x^3)$

(k) $y = (x^2 + 3)^{1/2}$,　　(l) $y = (x+1)/(x-1)$

4.2 Two of the derivatives obtained in problem (4.1) should be identical. Why?

4.3 Each of the following functions displays one minimum and one maximum. Identify both:

(a) $y = 7 + x^2 - 8x + \ln(x^6)$,　(b) $y = 15 - x - 1/(x-5)$

4.4 Michaelis and Menten analysed their data in terms of the equation that bears their names, $v = Va/(K_m + a)$, but they did not plot the rate v against the substrate concentration a. Instead, they plotted v against $\log a$, assuming that this plot has a point of inflection at the point where $a = K_m$ and that the maximum slope of the plot is $0.576V$. Prove that their assumptions are correct.

4.5 Differentiate the equation $v = Va/(K_m + a)$ with respect to a. What are the values of $\dfrac{dv}{da}$ when (a) $a = 0$; (b) $a = K_m$; (c) a approaches infinity?

4.6 Consider a pH profile defined by the following equation, in which h represents the hydrogen-ion concentration $[H^+]$ and y is a constant:

$$y = \frac{\tilde{y}}{\left(1 + \dfrac{h}{K_1} + \dfrac{K_2}{h}\right)}$$

(a) At what value of h is y a maximum?

(b) Show that this value of h corresponds to a pH of $(pK_1 + pK_2)/2$.

4.7 Under certain limiting conditions, the binding of a small ligand to a protein can be approximated by an equation of the following form:

$$Y = \frac{Kx^h}{1 + Kx^h}$$

in which Y is a measure of the extent of binding, x is the concentration of ligand, and K and h are constants. Putting $K = 1$ (arbitrary unit) and $h = 2$, sketch the form of curve given by this equation for a plot of Y against x (for positive x).

4.8 For the equation given in problem (4.7), what would be the slope of a plot of $\log[Y/(1-Y)]$ against $\log x$?

4.9 Differentiate the following equation with respect to t, treating p and t as variables and V, K_m and a_0 as constants:

$$Vt = p + K_m \ln \frac{a_0}{a_0 - p}$$

CHAPTER 5

Integral calculus

5.1 **Increases in area**

Figure 5.1 represents an arbitrary function y of x, and the area of the region bounded by the curve, the x-axis and the verticals $x = x_0$ and $x = x_1$ is designated A. For the moment we have no way of knowing the value of A, but first let us ask a simpler question: what is the *increase* δA in this area if x_1 is increased to $x_1 + \delta x_1$? Apart from the small approximately triangular area that is shaded in the diagram, all of the increase in area is accounted for by the tall thin rectangle of height y_1 and width δx, which has area $y_1 \, \delta x$. In symbols,

$$\delta A \approx y_1 \, \delta x$$

If δx is made smaller, the small shaded triangle decreases rapidly in area, not only absolutely, but also as a fraction of the area of the rectangle. Consequently, the approximation embodied by the above equation is very accurate if δx is very small, and in the limit as δx approaches zero it becomes exact. This is not very helpful, however, if we want to know the value of δA when δx is large, but the *integral calculus* provides us with tools for overcoming the difficulty. We begin by noting that if we divide both sides of the equation by δx we have

$$\frac{\delta A}{\delta x} \approx y_1$$

This sort of equation should be familiar from the discussion in Chapter 4, and in the limit it is exact:

$$\frac{\mathrm{d}A}{\mathrm{d}x} = y_1$$

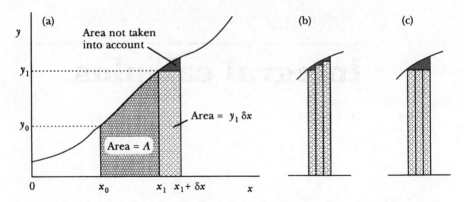

Fig. 5.1 The area under a curve. (**a**) The increase in A as x increases from x_1 to $x_1 + \delta x$ is approximately the area $y_1 \delta x$ of the rectangle obtained by treating $y = y_1$ as a constant over the range considered, but this leaves the shaded area out of account. (**b**) The approximation can be improved by treating the change as the sum of several small changes, recalculating y at the end of each small change. (**c**) However, if y is not recalculating each time this is no better than making one large change.

This expresses the important generalization that the derivative of the area between the x-axis and any curve $y = f(x)$ is equal to the ordinate variable y_1 calculated at the particular abscissa value x_1 to which the derivative applies. As we shall see, this relationship means that we can calculate the area under any curve whose algebraic expression can be recognized as the derivative of some other function. First, however, we must see how a large area can be treated as the sum of many small areas.

Figure 5.1(b) shows that when the increment δx is made smaller, the area not taken into account by the calculation decreases much more rapidly than the decrease in the area that is taken into account; thus, we can calculate the increase in area more accurately if we treat it as the sum of several strips narrower than the original one. However, this only works if we recalculate y_1 each time. As Figure 5(c) illustrates, we are no better off with three narrow strips than with one wide one if y_1 is not recalculated for each strip.

As long as δx is small enough for the relationship $\delta A \approx y_1 \delta x$ to be a good approximation we can write

$$\Delta A \approx \sum_{}^{n} \delta A = \sum_{}^{n} y \, \delta x$$

for the increase in area ΔA brought about by increasing x by $\Delta x = n \, \delta x$, i.e. in n small increments of δx. At first sight, this is easy to calculate, as $\sum_{}^{n} \delta x$ is just $n \delta x$, or Δx, but we cannot take y outside the summation, as this would be equivalent to treating it as a constant whereas as we have seen in Fig. 5.1(c) it has to be recalculated after each increase in x.

If the approximation is made better and better by making δx smaller and smaller, the number of increments n must be increased concomitantly to keep Δx the same size. In the limit, the relationship is exact when there are an infinite number of infinitesimal increments:

$$\Delta A = \lim_{n \to \infty, \delta x \to 0} \sum^{n} y \, \delta x = \int y \, dx$$

The new symbol $\int$ is used to represent this sum of an infinite number of terms, i.e. the middle expression is a definition of the right-hand expression. We call this an *integral* and the process of calculating it is called *integration*.

Because of the relationship $\frac{dA}{dx} = y_1$ referred to earlier, we might expect to be able to integrate any function by reversing the rules for differentiation, but in practice it is often difficult or impossible to recognize the function of interest as the result of differentiating some other function, and there is nothing in the integral calculus that corresponds to differentiating from first principles, i.e. there is no general method that is guaranteed to work.

Nonetheless, some simple functions are easily recognized as derivatives of other functions. For example, in

$$y = 3x^2 + 2x + 5$$

each term is easily recognized as the result of differentiating a power of x with respect to x: $3x^2$ is what we would get by differentiating x^3, $2x$ is the result of differentiating x^2, 5 is the result of differentiating $5x$, and 0 is the result of differentiating any constant α. This last term is perhaps surprising, but it is important: differentiating any constant produces a zero, and so when we apply the process in reverse we must not forget to account for a zero term in the sum. Taking all this into account it is easy to write down the integral of the above expression:

$$\int y \, dx = x^3 + x^2 + 5x + \alpha$$

5.2 **Definite and indefinite integrals**

We have seen that integration can be regarded as the inverse of differentiation. Just as the derivative of any constant is zero, therefore, the integral of zero is any constant and the constant α that appeared in the integration at the end of the previous section can have any value. A constant of this sort is called a *constant of integration*. Its meaning, and the reason for its appearance, can be seen by reflecting that in the previous section we discussed the increase in an area A without specifying what value of A we had to start with. It follows that integration is *ambiguous*: although differentiation leads to a unique result this result could have been the result of differentiating an infinite number of other functions.

Example 5.1 Integrating a polynomial

Integrate the function $y = 4x^2 - 2x + 5$, and evaluate the constant of integration by incorporating the information that the integral has a value of 5 when $x = 1$.

The integral is the sum of the integrals of the three terms (increase each index by 1 and divide the coefficient by the new index), plus a constant:

$$\int y \, dx = 2x^3 - x^2 + 5x + \alpha$$

The numerical information given requires that

$$5 = 2 - 1 + 5 + \alpha$$

and hence $\alpha = -1$.

The ambiguity implied by the need for a constant of integration can be resolved by specifying *limits* between which the integration is to be done. Thus, an expression such as

$$\int (3x^2 + 2x + 5) \, dx$$

is ambiguous because although it tells us to add together terms of the form $(3x^2 + 2x + 5) \, dx$, it does not tell us where to begin and end. This sort of integral is called *an indefinite integral*. However, if we wish to start at $x = 1$ and finish at $x = 2$, we can indicate this by writing the integral as a *definite integral*:

$$\int_1^2 (3x^2 + 2x + 5) \, dx$$

where the numbers 1 and 2 written beside the integration symbol define the *limits* of the definite integral, which can be regarded as the difference between the values of the indefinite integral at these limits:

$$\int (3x^2 + 2x + 5) \, dx = x^3 + x^2 + 5x + \alpha$$

$$= \begin{cases} 7 + \alpha & \text{at } x = 1 \\ 22 + \alpha & \text{at } x = 2 \end{cases}$$

On subtracting one from the other, the ambiguity disappears because, regardless of the value of α, it vanishes from the difference:

$$\int_1^2 (3x^2 + 2x + 5) \, dx = [x^3 + x^2 + 5x + \alpha]_1^2 = 22 - 7 = 15$$

It is sometimes convenient, as in this example, to indicate the limits outside square brackets after the form of the indefinite integral has been decided. An expression of this sort can be read as an instruction to 'evaluate the expression contained in the brackets at both limits and subtract one from the other'.

....

Example 5.2 Evaluating a definite integral

Evaluate $\int_{-1}^{1} (9x^2 - 2x - 5)\,dx$.

The indefinite integral is $3x^3 - x^2 - 5x + \alpha$, which has values of $-3 - 1 + 5 + \alpha = 1 + \alpha$ when $x = -1$ and $3 - 1 - 5 + \alpha = -3 + \alpha$ when $x = 1$. The definite integral is the difference between these two, i.e.

$$[3x^3 - x^2 - 5x]_{-1}^{1} = -3 - 1 = -4$$

The constant of integration can always be omitted when evaluating a definite integral, because it cancels. However, it is best to include it if omitting it encourages you to omit it in places where it omission is incorrect.

....

Notice that an indefinite integral is a *function* that can have any value, not only because of the undefined constant, but also because it can be evaluated at any value of x. By contrast, a definite integral defines a *number*. It follows that evaluation of a definite integral requires two stages: first, determination of the function corresponding to the indefinite integral; and second, evaluation of this function at the specified limits. The second stage is usually much easier than the first: given the expression for an indefinite integral, we can always evaluate it, but it is not unusual to begin with a function that we cannot integrate at all.

A chemical example is the rate of a simple first-order reaction, which can be expressed as

$$\frac{da}{dt} = -ka$$

where a is the concentration of the reacting substance at time t, k is a constant, and the minus sign is to show that the reacting substance disappears, i.e. its concentration decreases, as the reaction proceeds. We can rearrange this to show that an infinitesimal change da in a is related to an infinitesimal change dt in t as follows:

$$\frac{da}{a} = -k\,dt$$

Suppose we want to know the total decrease in concentration that occurs between $t = 0$ and $t = t_1$; we can find this by integrating:

$$\int_{t=0}^{t=t_1} \frac{da}{a} = -k \int_{0}^{t_1} dt$$

We write the limits on the left-hand side explicitly as $t = 0$ and $t = t_1$ (rather than just 0 and t_1, respectively) because the expression after the integral sign is written in terms of a different variable, a. Although just writing 0 and t_1, respectively, would probably be understood, it is safer to be explicit. On the right-hand side, limits 0 and t_1 are unambiguous because the variable on this side is t.

In this case the indefinite integrals are straightforward: on the left-hand side, we know that differentiating $\ln x$ produces $1/x$, so we should be able to deduce that integrating da/a will produce in a (plus a constant); on the right-hand side, we know that differentiating x produces 1, so integrating 1 produces x (plus a constant). So:

$$[\ln a]_{t=0}^{t=t_1} = -k[t]_0^{t_1}$$

If we define $a = a_0$ at $t = 0$ and $a = a_1$ at $t = t_1$, this becomes

$$\ln a_1 - \ln a_0 = -kt_1$$

Remembering what it means to subtract one logarithm from another (Section 3.5), we can write this as

$$\ln\left(\frac{a_1}{a_0}\right) = -kt_1$$

and then take exponentials of both sides and rearrange it into an expression for the concentration at any time:

$$a_1 = a_0 \exp(-kt_1)$$

This derivation has been pedantic in that different symbols t_1 and t were used with different meanings—t was a variable whereas t_1, a constant, was a particular value of t, though many authors would use the same symbol for both, taking the distinction to be obvious. Thus, the modern practice in chemistry is to normally write the above result as

$$a = a_0 \exp(-kt)$$

Using the same symbol with both meanings does no harm provided one realizes what has been done and one is willing to tolerate expressions such as $\int_0^t dt$ in which the same symbol t is used with two different meanings. It is especially convenient in applications where the upper limit is to be treated as a variable in subsequent analysis.

In kinetic applications, it is often clearer not to write definite integrals at all but to work with indefinite integrals with constants of integration that can be defined or evaluated. Therefore, we might write the initial equation for the first-order reaction as

$$\int \frac{da}{a} = -k \int dt$$

and integrate it to give

$$\ln a = -kt + \alpha$$

Notice that the same constant of integration α serves both integrals: if we had included constants of integration on both sides, we could subtract one from the other to get the constant we have called α. If we define $a = a_0$ at $t = 0$ then substitution of these two values into the above equation gives the value of α,

$$\ln a_0 = \alpha$$

which can then be substituted into the equation:

$$\ln a = -kt + \ln a_0$$

Subtraction of $\ln a_0$ from both sides, combining the two logarithmic terms into one and taking exponentials gives the same result as before:

$$a = a_0 \exp(-kt)$$

but without any confusion about whether a and t are being treated as variables or constants.

It is important to note that the constant of integration in this example was not zero, even though it was evaluated at $t = 0$. Although a few elementary kinds of functions, most notably polynomials of the form $A + Bx + Cx^2 + Dx^3 + \cdots$, do give a zero value for the integrated function which has a value of zero when the lower limit is zero, this is *not* a general rule. Polynomial functions may be common in elementary textbook accounts of integration but they are not common in scientific applications of integration and one must not suppose that there is any kind of general rule that allows constants of integration to be ignored. In kinetics—which accounts for most of the integrating that a biochemist is likely to do—it is *never* safe to ignore constants of integration or to assume that integrals are zero at zero time.

5.3 **Simple integrals**

In principle, any function can be differentiated, but the converse is by no means true. The only functions that are easy to integrate are those that are recognizable as derivatives of known functions. For example,

$$\int A \, dx = A x + \alpha$$

$$\int A x \, dx = \frac{1}{2} A x^2 + \alpha$$

$$\int A x^2 \, dx = \frac{1}{3} A x^3 + \alpha$$

or in general,

$$\int A x^i dx = \frac{A x^{i+1}}{i+1} + \alpha$$

Terms of this kind are called *polynomial terms*, and the rule for integrating them is as follows: increase the index by 1, divide by the *new* index, and add a constant.

The integral of a sum (or difference) is also straightforward: if u and v are functions of x then

$$\int (u+v)\,dx = \int u\,dx + \int v\,dx$$

As we already know how to integrate a polynomial term, this immediately tells us how to integrate a complete polynomial: we just integrate each term separately and add all the results together. So, for example,

$$\int (3+2x-6x^2)\,dx = \int 3\,dx + \int 2x\,dx - \int 6x^2\,dx = 3x + x^2 - 2x^3 + \alpha$$

As before, we only need to write one constant, because the sum of three constants is still a constant.

Two other simple functions occurred as derivatives in Chapter 4 and are consequently easy to integrate. Integrating a variable to the power -1 is just the reverse of differentiating a natural logarithm, so

$$\int \frac{dx}{x} = \ln x + \alpha$$

and as differentiating $\exp(x)$ leaves it unchanged, it should be obvious that integrating it also leaves it unchanged apart from the obligatory constant:

$$\int \exp(x)\,dx = \exp(x) + \alpha$$

5.4 **Other common integrals**

Although rules exist for integrating products and other functions, these are by no means as easy to apply as the rules for differentiating the same sorts of function. The need for them occurs so infrequently in elementary biochemistry that they are best dealt with as special cases when they do occur, in other words by looking them up in standard tables (see below). There are, however, a few functions that occur so often in kinetics that it is useful to be able to recognize them. The first is the following:

$$\int \frac{dx}{A+Bx} = \frac{1}{B}\ln(A+Bx) + \alpha$$

It is easy to confirm that this is correct by differentiating the right-hand side as a function of a function. It is less obvious how to deal with *products* of functions of the same kind, which also occur quite often in kinetics. For example, we may be faced with an integral of the following kind:

$$\int \frac{dx}{(a_1+b_1x)(a_2+b_2x)}$$

The essential point here is to realize that a fraction with a denominator that consists of a product of polynomials can always be written as a sum of separate fractions, in other words we can always find values of A_1 and A_2 such that

$$\frac{1}{(a_1 + b_1 x)(a_2 + b_2 x)} \equiv \frac{A_1}{a_1 + b_1 x} + \frac{B_2}{a_2 + b_2 x}$$

....

Example 5.3 Partial fractions

Determine values of P and Q such that

$$\frac{P}{5 + 2x} + \frac{Q}{5 - 2x} \equiv \frac{7}{25 - 4x^2}$$

Note first that the denominator on the right-hand side is a difference between two squares, and is thus the product of the denominators of the two fractions on the left-hand side. Cross-multiplying, therefore,

$$(5 - 2x)P + (5 + 2x)Q \equiv 7$$

As this is an identity it is true for any values of x, so choosing $x = 0$ and $x = 1$ we have

$$5P + 5Q = 7$$
$$3P + 7Q = 7$$

Subtracting, we have $2P - 2Q = 0$, so $P + Q$, hence $10P = 7$, so $P = Q = 0.7$.

We are not obliged to choose $x = 0$ and $x = 1$ to evaluate the identity. We can choose any values we like (as long as they are different), but it makes sense to choose ones that are easy to calculate with.

....

The *identity sign* ($\equiv$) is used here to emphasize that the relationship must apply *regardless* of the value of x, i.e. this is not an equation that we might solve to determine the value of x. The two separate fractions produced in this way are called *partial fractions*. Multiplying both sides of the identity by the denominator of the left-hand side gives a simpler identity that can be used to determine the values of A_1 and A_2 that will make it true:

$$1 \equiv A_1(a_2 + b_2 x) + A_2(a_1 + b_1 x)$$

As this must be true for any value of x, it must apply when $x = 0$ and also when $x = 1$, so we can substitute these values of x,

$$1 = A_1 a_2 + A_2 a_1$$

$$1 = A_1(a_2 + b_2) + A_2(a_1 + b_1)$$

$$= A_1 a_2 + A_1 b_2 + A_2 a_1 + A_2 b_1$$

Subtraction of the first equation from the second gives

$$0 = A_1 b_2 + A_2 b_1$$

from which it is obvious from inspection that $A_2 = -A_1 b_2 / b_1$, and substitution of this into the first equation gives the expressions for the two constants:

$$A_1 = \frac{b_1}{a_2 b_1 - a_1 b_2}, \qquad A_2 = \frac{-b_2}{a_2 b_1 - a_1 b_2}$$

The original fraction can thus be written as the sum of two fractions, which can be integrated separately:

$$\int \frac{dx}{(a_1 + b_1 x)(a_2 + b_2 x)} = \int \frac{A_1 dx}{a_1 + b_1 x} + \int \frac{A_2 dx}{a_2 + b_2 x}$$

$$= \frac{A_1}{b_1} \ln (a_1 + b_1 x) + \frac{A_2}{b_2} \ln (a_2 + b_2 x) + \alpha$$

Although the derivation of the values of A_1 and A_2 was just an exercise in algebra and thus not essential to understanding the principles of integration, it is useful to notice that the two coefficients are the same apart from their signs, and so the two terms can be combined by using the relationship for subtracting one logarithm from another:

$$\int \frac{dx}{(a_1 + b_1 x)(a_2 + b_2 x)} = \frac{1}{a_2 b_1 - a_1 b_2} \ln \left(\frac{a_1 + b_1 x}{a_2 + b_2 x} \right) + \alpha$$

This is the form of the result that one is likely to find in a kinetics textbook or in a table of integrals, so it is important to understand that it is equivalent to the result that we have derived.

Another integral that is often needed in kinetic problems is a fraction with a numerator that is proportional to x and a denominator that is a linear function of x:

$$\int \frac{x \, dx}{A + Bx}$$

This is the simplest example of a general class of problems that are most easily handled by defining a new variable u equivalent to the denominator and then writing the whole fraction in terms of it:

$$u = A + Bx$$

and so

$$x = \frac{u - A}{B}$$

$$\frac{du}{dx} = B, \quad \text{or} \quad du = B \, dx$$

Now, although the complicated expression for x in terms of u may seem to nullify the advantage of simplifying the denominator by writing it as u, in fact it is much easier to handle a fraction with a complicated numerator and a simple denominator than the reverse. Substitution of these expressions into the original gives

$$\int \frac{x\,dx}{A+Bx} = \frac{1}{B^2}\int \frac{(u-A)\,du}{u}$$

$$= \frac{1}{B^2}\left(\int du - A\int \frac{du}{u}\right)$$

$$= \frac{u - A\ln u}{B^2} + \alpha$$

The solution in terms of the original variable x then follows by converting u back into $A+Bx$:

$$\int \frac{x\,dx}{A+Bx} = \frac{A+Bx-A\ln(A+Bx)}{B^2} + \alpha$$

This result is not of sufficient importance in itself to be worth memorizing, but the method by which it was obtained is often useful in integration problems: the general approach is to search for a suitable function u that allows an easier integration.

The main results of this and the preceding section are given in Table 5.1.

Table 5.1 Common integrals.

$$\int Ax^i\,dx = \frac{Ax^{i+1}}{i+1} + \alpha \quad \text{(for any value of } i \text{ except } -1)$$

$$\int \frac{dx}{x} = \ln x + \alpha$$

$$\int (u+v)\,dx = \int u\,dx + \int v\,dx$$

$$\int \exp(Ax)\,dx = \frac{1}{A}\exp(Ax) + \alpha$$

$$\int \frac{dx}{A+Bx} = \frac{1}{B}\ln(A+Bx) + \alpha$$

$$\int \frac{x\,dx}{A+Bx} = \frac{A+Bx-A\ln(A+Bx)}{B^2} + \alpha$$

In all cases x is a variable, A, B and i are constants, u and v are functions of x, and α is a constant of integration. All of the integrals in this table apart from the last are worth memorizing.

Box 5.1 **Integrating an unknown function**

In this explanation, we assume that the unknown function is a function of the variable x and we need to integrate it with respect to x, A, B, and i are constants.

1   If it consists of a sum of separate terms then try to integrate each term separately.

2   If any such term is of the form Ax^i then increase the index by 1 and divide by the new index: $Ax^{i+1}/(i+1)$.

3   If not, but it is of the form $A\ln(Bx)$ then replace it with A/x. (B, the coefficient of x, can be ignored: can you see why?)

4   If not, but it is of the form $A\exp(Bx)$ then just divide by the coefficient of x: $(A/B)\ln(Bx)$

5   If not, then if there is a way of defining a function $u=f(x)$ that can be integrated with respect to u by following the above rules, integrate this and substitute back into terms of x afterwards.

6   If all of the above fails, then seek help! Remember that not all functions can be integrated. You may be struggling with an impossible problem.

7   The final integral is the sum of the individual integrals plus a constant.

8   *Never forget the constant!* Remember that if you have obtained the answer from a set of tables, the constant may well have been omitted to save space: compilers of such tables usually assume that users are sophisticated enough to know that a constant has to be added.

9   If you like, and there is a logarithmic term in the answer, you can incorporate the constant in the logarithmic term rather than writing it separately: $\ln x+\alpha$ is the same as $\ln(Ax)$ if A is defined so that $\ln Ax=\alpha$. Although it is common practice to use a greek letter (usually α) for a constant of integration written as a separate term and a capital roman letter, such as A, for one that is incorporated in a logarithmic term, this is not of course compulsory.

5.5 **Integral of** $1/x$

The following pair of relationships:

$$\frac{\mathrm{d}}{\mathrm{d}x}(\ln x)=\frac{1}{x}, \qquad \int \frac{\mathrm{d}x}{x}=\ln x+\alpha$$

have been referred to several times in this chapter and the previous one, but have not been shown to be correct. A rigorous proof is hardly essential in an elementary course but it is still useful to invoke plausibility arguments to be satisfied that they are reasonable, at least.

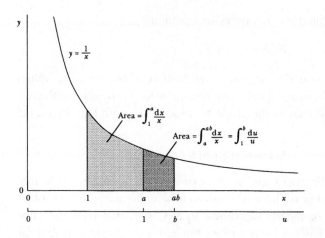

Fig. 5.2 Plot of the hyperbola $y = 1/x$. The u-axis is an alternative abscissa axis defined by $x = ua$ or $x = u/a$. The area under the curve between $x = a$ and $x = ab$ is not changed in magnitude by calling it the area under the curve between $u = 1$ and $u = b$.

Figure 5.2 represents the function $y = 1/x$. Let us now suppose that a function $F(a)$ exists that represents the definite integral between 1 and some number a:

$$\int_1^a \frac{dx}{x} = F(a)$$

For the moment, we specify nothing about the nature of this function $F(a)$; by the usual interpretation of a definite integral, it defines the area bounded by the curve, the x-axis, and the vertical lines $x = 1$ and $x = a$. The corresponding area between $x = 1$ and $x = ab$ can be expressed similarly:

$$\int_1^{ab} \frac{dx}{x} = F(ab)$$

This is, however, the region obtained by combining the area from $x = 1$ to $x = a$ with that from $x = a$ to $x = ab$, so its area can equally well be written as the sum of the two separate areas:

$$\int_1^{ab} \frac{dx}{x} = F(ab) = \int_1^a \frac{dx}{x} + \int_a^{ab} \frac{dx}{x} = F(a) + \int_a^{ab} \frac{dx}{x}$$

Note that for the moment we have no way of writing the integral from a to ab as an instance of the function F. However, if we now define a new variable u such that $u = x/a$, then $u = 1$ when $x = a$, and the right-hand term can be written in terms of u as follows:

$$\int_a^{ab} \frac{dx}{x} = \int_1^b \frac{du}{u}$$

However, the definition of F implied nothing about what variable it was acting on, so it applies just as well to u as to a, and so

$$\int_1^b \frac{du}{u} = F(b)$$

which can now be substituted into the previous equation:

$$F(ab) = F(a) + F(b)$$

and we see that the function $F(x)$ has the property that adding two $F(x)$ values together is the same as calculating the the function value for the product of the two x values. This, however, is exactly the property we associate with logarithms, i.e. it is equally true that

$$\ln(ab) = \ln a + \ln b$$

It is thus at least a possibility that F and ln *are the same function*. It happens to be true and not just a possibility, but we have not proved it here, because there are other functions, such as $\log_{10}$, or indeed logarithms to any base, that have the same property. The aim here has not been a proof but just a demonstration that the standard result is plausible, i.e. that

$$\int \frac{dx}{x} = \ln x + \alpha$$

The logarithm of a constant is another constant. It is legitimate, therefore, and often convenient, to express α as the logarithm of a different constant, A, and incorporate it into the first logarithmic term:

$$\int \frac{dx}{x} = \ln x + \ln A = \ln Ax$$

Constants of integration are often written in this way, and so one should not be surprised to see results of integration written with no constant added to the whole expression but instead with a constant factor appearing within the logarithmic term.

5.6 Differential equations

One of the commonest reasons why one needs to integrate expressions in science is that scientific laws frequently make predictions about the derivatives of functions of interest rather than about the functions themselves. Consider, for example, a reaction between two molecules A and B:

$$A + B \rightarrow P$$

Elementary kinetic considerations suggest that, if the mechanism is simple, the rate of reaction is likely to be proportional both to the concentration of A and to the concentration of B, i.e.

$$\frac{dp}{dt} = kab$$

where a, b, and p are the concentrations of A, B, and P, respectively, at time t and k is a constant. Although this relationship is easy to arrive at and easy to understand, it does not tell us very much as it stands; in particular, it does not tell us how much product has been formed after a given amount of time.

To convert it into a more useful expression, we must first reduce the number of variables, because it contains four variables (a, b, p, and t) and cannot be integrated in this form. However, two of the four variables can be eliminated by taking account of the stoichiometric constraint that every molecule of P that appears is the result of the disappearance of one molecule of A and one molecule of B; it follows that $(a+p)$ and $(b+p)$ are constants. If a_0 and b_0 are defined as the values of a and b, respectively, when $t=0$ and $p=0$, then a and b can be written as (a_0-p) and (b_0-p), respectively:

$$\frac{dp}{dt} = k(a_0-p)(b_0-p)$$

This is an example of a *differential equation*. This name suggests that it belongs in the domain of differential calculus, but manipulation of differential equations actually requires knowledge of integral calculus, because they have to be integrated to remove the derivative. As it stands, both variables in the example appear on the left-hand side of the equation, but it is easy to *separate* them so that each side of the equation contains only one kind of variable:

$$\int \frac{dp}{(a_0-p)(b_0-p)} = \int k\,dt$$

Integrating the right-hand side is trivial, and the left-hand side can be integrated by partial fractions (Section 5.4):

$$-\frac{\ln(a_0-p)}{a_0-b_0} + \frac{\ln(b_0-p)}{a_0-b_0} = kt + \alpha$$

Putting $p=0$ when $t=0$ allows the constant of integration α to be evaluated:

$$-\frac{\ln a_0}{a_0-b_0} + \frac{\ln b_0}{a_0-b_0} = \alpha$$

When this is substituted into the previous equation, all the logarithmic terms have the same coefficient (apart from signs), so they can all be combined into one term:

$$\ln\left[\frac{a_0(b_0-p)}{b_0(a_0-p)}\right] = (a_0-b_0)kt$$

or

$$\frac{a_0(b_0 - p)}{b_0(a_0 - p)} = \exp[(a_0 - b_0)kt]$$

This contains no derivatives, and is the solution of the original differential equation.

Not all differential equations permit the variables to be separated in this way. Ones that do not are not common in elementary biochemistry, however, and so it is not worthwhile devoting much study to methods of solving them. Unless one plans to be a specialist in a branch of biochemistry that requires differential equations to be solved frequently, it is much more efficient to seek expert help rather than struggle with problems that may have no solutions. It is always wise to remember that even quite simple kinetic models can lead to intractable mathematics and there is little point in wasting time over an impossible problem. For example, the simplest model commonly discussed in enzyme catalysis is the Michaelis–Menten mechanism:

$$\mathrm{E} \; + \; \mathrm{A} \; \underset{k_{-1}}{\overset{k_1}{\rightleftharpoons}} \; \mathrm{EA} \; \overset{k_2}{\longrightarrow} \; \mathrm{E} + \mathrm{P}$$

$$e_0 - x \qquad a \qquad\qquad x \qquad\qquad p$$

With concentrations defined as shown under the equations, application of elementary kinetic laws leads, without difficulty, to the following pair of simultaneous differential equations:

$$\frac{dx}{dt} = k_1(e_0 - x)(a_0 - x - p) - k_{-1}x - k_2x + k_{-2}(e_0 - x)p$$

$$\frac{dp}{dt} = k_2x - k_{-2}(e_0 - x)p$$

It is not difficult to remove x from these equations: first, we use the second equation to express x in terms of p, and differentiating this gives an expression for $\frac{dx}{dt}$ in terms of p and its derivatives; these two can then be substituted into the first equation to give a single differential equation containing only the two variables p and t. However, this is as far as we can go, because the resulting differential equation has no known solution.

If one arrives at an equation that is impossible to solve, there is little point in relying on one's mathematical knowledge and ability. Instead, one has the choice between trying to solve it *numerically* rather than algebraically, an approach that is largely outside the scope of this book, and restricting the problem so as to make it soluble. This latter approach is very important and it demands scientific rather than mathematical knowledge. Thus, a mathematician might point out that the problem discussed above becomes soluble if $k_1 = k_{-2}$ but the biochemist would most likely feel that as there was no particular reason for this special case to be true its solution was not of interest; it would be preferable to look for a way of restricting the problem that could be *made to be true* by suitable choice of experimental conditions. By far, the most commonly invoked restriction in enzyme kinetics is to require a_0 to be much larger than e_0 and to consider only the time scale in which $\frac{dx}{dt}$ can be regarded as negligible. This is the basic restriction used in *steady-state kinetics*, but it is not the only possible one. We could, for example, continue to assume $a_0 \gg e_0$ but consider a different time scale, in which p is negligible; this is the *pre-steady-state* condition. Or, less commonly, we could assume $e_0 \gg a_0$, the *single-turnover* condition. The purpose in mentioning these, however, is not to derive any kinetic results, but to emphasize that the decisions to be made are scientific, not mathematical: one needs some knowledge of enzymes and techniques for studying them to know that it is usually easy and convenient to achieve steady-state conditions, but that assuming $k_1 = k_{-2}$ is unlikely to be useful. The essential point is that the experimenter has control over the experimental conditions, but has no control over the constants that define the system.

To summarize this section, the most efficient approach to differential equations in biochemistry is to proceed as follows: first, check whether the variables can be decreased to two that can be separated on to the two sides of the equation. If so, the equation can in principle be solved by integrating the two sides. If this cannot be done immediately then tables of integrals or expert advice are likely to prove much more rapid than blind struggling. If the variables cannot be separated or the problem appears impossible for some other reason, then try to introduce biochemically sensible restrictions or approximations that allow the simplified equation to be solved. If all else fails then seek expert help, remembering that a mathematician will probably know much less biochemistry than you do and will need some guidance over matters that you regard as obvious.

Finally, a word about tables of integrals and differential equations. Tables of integrals are easy to use, provided you remember two things: first, constants of integration are usually omitted, on the assumption that the user knows enough mathematics to supply a constant of integration without constant reminders; second, the symbol log usually means ln, not, as is the practice in biochemistry, $\log_{10}$. Tables of differential equations are much more difficult to use as they commonly assume some understanding of the principles of solving differential equations.

5.7 Numerical integration: evaluating the area under a curve

It sometimes happens that we need to know the area under a curve but cannot do the necessary integration exactly, either because we do not have an algebraic expression for the curve or because we cannot integrate it. For example, the curve in Fig. 5.3 might represent the absorbance of the effluent from a chromatographic column as a function of the volume: if the absorbance of a sample is proportional to the concentration of some substance of interest, the total amount of the substance is proportional to the area under the curve. If it is not possible or convenient to integrate the curve analytically, the simplest way of estimating the area is to divide it into n parallel strips as shown and replace the short arcs between the ends of the strips by straight-line segments. It is then a matter of simple geometry to determine the areas of the trapezia produced. For example, between $x_0 = 0$ and $x_1 = h$, the area of the rectangle below the curve is $y_0 h$ and the area of the triangle between the rectangle and the curve is $(y_1 - y_0)h/2$, so the area of the whole trapezium is $(y_1 + y_0)h/2$. Similarly, the area of the adjacent trapezium is $(y_1 + y_2)h/2$, and so on. The total area A under all the straight-line segments is given by

$$A = [(y_0 + y_1) + (y_1 + y_2) + (y_2 + y_3) + \cdots + (y_{n-1} + y_n)]\frac{h}{2}$$

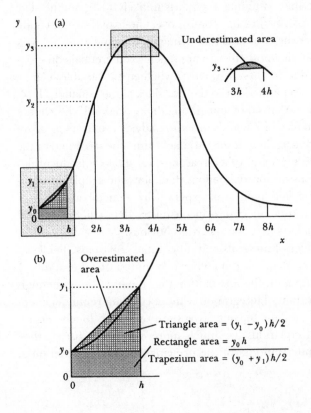

Fig. 5.3 The trapezium method for estimating the area under a curve. (**a**) The area is treated as a series of strips, and (**b**) the area of each of these is estimated as the area of a trapezium. When the second derivative of the function is positive, this overestimates the true area, as it includes a region above the curve. However, as illustrated in the insert near the maximum of the curve, under estimation occurs when the second derivative is negative.

Every y, apart from the first and last, appears twice, so this may also be written as follows:

$$A = \left(\frac{y_0}{2} + y_1 + y_2 + y_3 + \cdots + y_{n-1} + \frac{y_n}{2} \right) h$$

This formula is simple to calculate, but it is not very accurate, because the area under the curve is not exactly the same as the sum of the areas under the straight-line segments. As illustrated in Fig. 5.3(b), the trapezium may include a significant amount of area that is not under the curve, with corresponding overestimation of the area under the curve, and as the inset in Fig. 5.3(a) illustrates significant under-estimation can also occur, in regions of downward curvature. The accuracy can be improved, in principle without limit, by increasing the number of strips (i.e. by making h smaller), but this makes the method more tedious because of the large amount of calculation needed.

A more accurate method involves assuming that over a limited range the curve can be approximated by a parabola. To replace the unknown function $y = f(x)$, or a known function that we do not know how to integrate, we write a quadratic equation:

$$y \approx a + bx + cx^2$$

If we treat the approximation as exact over a short range, and evaluate y at three x values, $x_0 = 0$, $x_1 = h$, and $x_2 = 2h$, then

$$y_0 = a$$
$$y_1 = a + bh + ch^2$$
$$y_2 = a + 2bh + 4ch^2$$

Eliminating a by subtraction, we have

$$y_1 - y_0 = bh + ch^2$$
$$y_2 - y_1 = bh + 3ch^2$$

and bh by further subtraction,

$$y_2 - 2y_1 + y_0 = 2ch^2$$

Hence

$$ch^2 = y_2 - 2y_1 + y_0$$

and

$$bh = y_1 - y_0 - ch^2 = 2y_1 - \frac{3}{2} y_0 - \frac{1}{2} y_2$$

The original quadratic expression is easily integrated:

$$\int_0^{2h} y\,dx = \int_0^{2h} (a+bx+cx^2)\,dx = \left[ax\frac{1}{2} + bx^2 + \frac{1}{3}cx^3 \right]_0^{2h}$$

and we can now substitute the values of a, bh, and ch^2 to get the area of the first two strips:

$$\int_0^{2h} y\,dx = 2ah + 2bh^2 + \frac{8}{3}ch^3 = (y_0 + 4y_1 + y_2)\frac{h}{3}$$

The area of the next two pairs of strips can be derived similarly:

$$\int_{2h}^{4h} y\,dx = (y_2 + 4y_3 + y_4)\frac{h}{3}$$

$$\int_{4h}^{6h} y\,dx = (y_4 + 4y_5 + y_6)\frac{h}{3}$$

and so on. The areas of all these strips have to be added together to give the total area A:

$$A = (y_0 + 4y_1 + y_2)\frac{h}{3} + (y_2 + 4y_3 + y_4)\frac{h}{3} + (y_4 + 4y_5 + y_6)\frac{h}{3} + \cdots$$

If the first three strips were all there were, the total would be

$$A = (y_0 + 4y_1 + 2y_2 + 4y_3 + 2y_4 + 4y_5 + y_6)\frac{h}{3}$$

but it should be obvious how this can be generalized to larger numbers of strips, say five:

$$A = (y_0 + 4y_1 + 2y_2 + 4y_3 + 2y_4 + 4y_5 + 2y_6 + 4y_7 + 2y_8 + 4y_9 + y_{10})\frac{h}{3}$$

Examining this, we see that (1) the first and last y values have coefficients of 1; (2) all the other y values with even subscripts have coefficients of 2; (3) all the y values with odd subscripts have coefficients of 4. It follows that for any number $n/2$ of strips the formula is as follows:

$$A = [y_0 + 4(y_1 + y_3 + \cdots + y_{n-1}) + 2(y_2 + y_4 + \cdots + y_{n-2}) + y_n]\frac{h}{3}$$

This formula is known as *Simpson's Rule*. It is hardly more difficult to apply than the trapezium method considered first, and it is much more accurate. Unless the curve is highly complex, the formula is adequately accurate if the number of strips is about 10, and in most experimental examples inaccuracies in measuring the y values are a more important source of error than the formula.

5.8 Problems

5.1 Integrate the following:

(a) $\int (x^2 + 3x + 1) \, dx$, (b) $\int \left(x + \dfrac{1}{x} \right) dx$

(c) $\int \dfrac{dx}{2 + 3x}$, (d) $\int \exp(-3t) \, dt$

(e) $\int \dfrac{3x \, dx}{2 + x}$

5.2 Evaluate the following definite integrals:

(a) $\int_0^5 3x \, dx$, (b) $\int_1^2 \dfrac{dx}{x}$

(c) $\int_{-1}^1 3x^2 \, dx$, (d) $\int_{-2}^2 x^3 \, dx$

5.3 Express the following equation in the form of partial fractions and use the result to evaluate $\int y \, dx$:

$$y(3x + 1)(x + 2) = 5$$

5.4 Integrate the following expression after first separating it into two integrals:

$$\int \frac{(A + Bx) \, dx}{C + Dx}$$

5.5 The Michaelis–Menten equation can be written in the following form:

$$\frac{dp}{dt} = \frac{Va}{K_m + a}$$

in which p and a are the concentrations of product and substrate, respectively, at time t, and V and K_m are constants. Defining $a = a_0$ and $p = 0$ at $t = 0$, use the one-to-one stoichiometry of the reaction to eliminate a and solve the resulting differential equation to get an expression for Vt in terms of p.

5.6 The equation considered in problem (5.5) takes no account of the possibility of inhibition by the accumulating product. Show that the following equation, in which K_p is a constant and the other symbols are as in problem (5.5), can be integrated to provide an equation of the same form as the solution to problem (5.5):

$$\frac{dp}{dt} = \frac{Va}{K_m\left(1+\dfrac{p}{K_p}\right)+a}$$

What would be the form of a plot of $t/\ln[a_0/(a_0-p)]$ against $p/\ln[a_0/(a_0-p)]$?

5.7 The relationship between the intensity of light, I, and the distance x it has passed through an absorbing solution can be expressed as follows:

$$\frac{dI}{dx} = -kcI$$

where k is a constant and c is the molar concentration of absorbing substance. Putting $I=I_0$ at $x=0$, integrate this expression to obtain an expression for I in terms of x. If the *molar absorbance A* is defined such that

$$\log(I_0/I) = Acx$$

what is the relationship between A and k?

5.8 In a spinning ultracentrifuge cell, the outward flow rate of the mass of solute m is given, ignoring diffusion, by

$$\frac{dm}{dt} = \frac{\omega^2 x(1-\bar{V}\rho)M_r ADC}{RT}$$

where t represents time, x the distance from the axis of rotation, M_r the relative molecular mass of the solute, A the cross-sectional area of the cell, D the diffusion coefficient, c the concentration of solute in g/L, and the other symbols represent constants that are either known or easily measurable. The effect of diffusion can be represented by

$$\frac{dm}{dt} = -AD\frac{dc}{dx}$$

After spinning for a long period, the system achieves equilibrium and becomes time independent. The net flow rate, given by the sum of the two expressions, is then zero at all points. Obtain an expression relating c and x at equilibrium and hence suggest a plot that would allow M_r to be conveniently measured.

5.9 The following values of y at various values of x show measurements of the height of a recorder trace in a chromatographic separation. Estimate the areas under the two peaks by using Simpson's rule to evaluate $\int_1^9 y\,dx$ and

$\int_9^{17} y \, dx$. In each case compare the results with four strips with those obtained with eight.

x	y	x	y	x	y
0	0.03	7	5.75	14	2.68
1	0.04	8	2.69	15	1.21
2	0.51	9	1.62	16	0.28
3	2.53	10	2.27	17	0.03
4	5.81	11	3.41	18	0.12
5	7.64	12	4.08	19	0.15
6	8.11	13	3.82	20	0.09

CHAPTER 6

Solving equations

6.1 **Linear equations in one unknown**

An equation such as

$$4 + x = 8 - 2x$$

contains only one unknown quantity, x, and is *linear*, because it contains no terms higher than the first power of x. Although it is easy to solve such an equation, it is still useful to examine the processes involved as they provide a simple illustration of the methods used for solving more difficult equations. The first essential step is to rearrange the equation so that terms of the same type, in this case constants and terms in x, are brought together. It is often convenient to collect all constants on the right-hand side and other terms on the left-hand side. This may be done by use of the principle that we can apply any operation we like to an equation (other than multiplying by zero or dividing by zero) provided that we apply it identically to both sides. So, for example, we can subtract 4 from both sides of the above equation and add $2x$ to both sides:

$$4 + x - 4 + 2x = 8 - 2x - 4 + 2x$$

and after collecting together terms of the same type this becomes

$$3x = 4$$

The operation of subtracting, for example, 4 from both sides is clearly equivalent to 'bringing it from one side of the equation to the other and changing its sign'. Although this is a convenient and proper way of carrying out the operation, one should realize that it is no more than an example of applying the same operation to both sides of an equation. Continuing with the above example, we may divide both sides of the last equation by 3 to obtain the solution:

$$x = \frac{4}{3} = 1.33 \quad \text{approx.}$$

Box 6.1 illustrates this procedure again for a different example.

Equations that do not seem to be linear at first sight can often be rearranged into linear form by application of the same methods. For example, the equation

$$\frac{8x}{5+2x} = 3$$

Box 6.1 **General approach to solving a linear equation**

We take a linear equation in one unknown as an example: $\qquad 14 - x = 5x + 2$

1. Subtract each term in x on the right-hand side from *both* sides: $\qquad 14 - x - 5x = 5x + 2 - 5x$

2. Subtract each constant (term without x) on the left-hand from *both* sides: $\qquad 14 - x - 5x - 14 = 5x + 2 - 5x - 14$

3. Gather terms in x together and constants together: $\qquad -x - 5x + 14 - 14 = 5x - 5x + 2 - 14$

4. Combine terms of the same sort: $\qquad -6x = -12$

5. Divide *both* sides by the coefficient of x: $\qquad x = 2$

Once you have a clear idea of what is going on, you can safely take some short cuts. In particular, subtracting the same term from both sides of an equation has exactly the same effect as moving a term from one side of the equation to the other and changing its sign:

Starting from the same equation: $\qquad 14 - x = 5x + 2$

1. Bring all terms in x to the left-hand side, changing the signs of any that are moved: $\qquad 14 - x - 5x = 2$

2. Bring all constants to the right-hand side, changing the signs of any that are moved: $\qquad -x - 5x = 2 - 14$

... and continue as before.

becomes linear after both sides of the equation are multiplied by $5 + 2x$:

$$8x = 15 + 6x$$

which can be solved in the same way as before to give $x = 7.5$.

The only difficulty that may arise in solving a simple linear equation is that it may turn out to be *singular*, which means that it contains less information than it seems to contain. Consider, for example, the following equation, which appears to be an ordinary linear equation:

$$5x + 2 - 3(x + 8) = 2x - 22$$

Multiplying out the bracket, however,

$$5x + 2 - 3x - 24 = 2x - 22$$

collecting terms,

$$5x - 3x - 2x = -22 - 2 + 24$$

taking out the common factor x,

$$(5 - 3 - 2)x = -22 - 2 + 24$$

and evaluating the numerical expressions, we get

$$0x = 0$$

which is true regardless of the value of x and, consequently, tells us nothing about the value of x. Examples as simple as this seem rather artificial and trivial, but singular equations can be a serious problem because they can appear in the last stages of solving a more complex equation or set of equations. This often happens in the analysis of experiments that have been badly planned so that they do not in fact provide the information they were intended to provide, and I shall consider some more realistic examples later in this chapter.

6.2 **Rearranging equations**

Equations in science often contain the information we require but in inverse form, i.e. instead of giving the required unknown in terms of known quantities an equation may express a known quantity in terms of the unknown. In such cases we can often arrive at a more useful formulation of the equation by solving it in the same way as we would if it were expressed numerically.

For example, one method of determining how tightly a small molecule binds to a protein involves allowing the small molecule to equilibrate across a membrane that is impermeable to the protein, which is confined to one side. At equilibrium, the free concentrations of ligand are the same on both sides, but the total concentration is not, because on the side containing the protein it includes the concentration of the protein–ligand complex as well as of the free ligand. The bound concentration a_{bound} (which is also the concentration of the protein–ligand complex) can then be found by subtraction. In an experiment of this kind the total ligand concentration can often be represented by an equation of the following form:

$$a_{bound} = \frac{P_0 \cdot a_{free}}{K + a_{free}}$$

in which P_0 is the total protein concentration, a_{free} is the free ligand concentration and K is the dissociation constant. In planning an experiment, however, we may wish to estimate in advance the free concentration that will produce a particular bound concentration for some assumed value of K. To find this, we need to rearrange the equation so that it expresses a_{free} in terms of a_{bound} rather than vice versa.

The problem is of the same kind as the second example considered in the previous section; it is just expressed in terms of algebraic symbols rather than numbers. The method of solving it is the same: we multiply by both sides of the equation by $K + a_{free}$:

$$a_{bound}(K + a_{free}) = P_0 \cdot a_{free}$$

collect terms in a_{free} on the left-hand side and other terms on the right-hand side:

$$a_{bound} \cdot a_{free} - P_0 \cdot a_{free} = -a_{bound} \cdot K$$

take out the factor a_{free},

$$a_{free}(a_{bound} - P_0) = -a_{bound} \cdot K$$

and divide by both sides by $a_{bound} - P_0$ to give the solution:

$$a_{free} = \frac{-a_{bound} \cdot K}{a_{bound} - P_0}$$

6.3 **Simultaneous linear equations**

If we have two or more unknown quantities a single equation is not sufficient to define their values. For example, the equation

$$2x + 3y = 12$$

can be satisfied by $x = 3$, $y = 2$, but this is not a unique solution and there are an infinite number of other possible solutions, such as $x = 4.5$, $y = 1$. In fact, if we apply the principles of the previous section to express y in terms of x:

$$y = (12 - 2x)/3$$

we see that we can assign any value whatsoever to x and still be able to find the appropriate value of y such that x and y satisfy the equation exactly. When there are fewer equations than unknowns, the problem is said to be *underdetermined*. For there to be a unique solution, the number of independent equations must be equal to the number of unknowns. The reason for this is that each equation can be used to express one unknown in terms of the others but, once that has been done, no further information is available from that equation. If an equation is lost each time an unknown is eliminated, it is clear that one can only finish with a single equation in a single unknown if the numbers of equations and unknowns were initially the same. Although the approach is general, I shall consider only the case of two equations in two unknowns, because it is rare in elementary biochemistry for one to have problems with three or more unknowns. Such equations are called *simultaneous equations* because they define two (or more) relationships that are true simultaneously.

Let us apply the approach indicated above to the following pair of equations:

$$2x + 3y = 12, \qquad 2x + y = 8$$

Treating the first equation as a linear equation in y with x as an unknown constant, we have

$$3y = 12 - 2x$$

Hence

$$y = 4 - \frac{2x}{3}$$

Substituting this for y in the second equation (the first being of no further use to us) we have

$$2x + 4 - \frac{2x}{3} = 8$$

which may be solved as a simple linear equation in x to yield $x = 3$, and, after substitution back into the expression for y, $y = 2$.

This approach is superficially attractive because it can be applied mechanically, i.e. without thought, and it always ought to work. Nonetheless, it is not the best way of solving simultaneous equations and one can often proceed more simply and with less chance of error by thinking first.

For solving the particular equations considered, there are at least two possible ways of improving the method. First, we ought to notice immediately just by looking at the two equations ('by inspection', in mathematical language) that the second equation gives a simpler expression for y than the first does, i.e.

$$y = 8 - 2x$$

and by substituting this into the first equation we can avoid fractions:

$$2x + 3(8 - 2x) = 12$$

etc., with the same solution.

Second, we should also notice from inspection that x is multiplied by the same constant 2 in both equations. Consequently, we can eliminate it immediately by subtracting the second equation from the first:

$$2x + 3y = 12$$
$$2x + \ y = \ 8$$
$$\overline{\ 2y = 4}$$

Hence $y = 2$, etc. The subtraction of one equation from another is justified by the fact that *if* $2x + y = 8$ then subtracting $2x + y$ is equivalent to subtracting 8 and so, when we subtract $2x + y$ from the left-hand side and 8 from the right-hand side, we are applying identical operations to the two sides of the equation.

This third approach is clearly the simplest and quickest of the three, but it seems to require the coincidence of finding the same term—$2x$ in this example—in both equations, and as we cannot expect this to happen very often we may feel that it is not a very useful approach in general. In fact, however, we can always ensure that such 'coincidences' occur by multiplying by appropriate factors in the first instance. For example, if

$$3x - 5y = 3, \qquad 2x + 3y = 21$$

then multiplying the first equation by 2, the coefficient of x in the second equation, and the second equation by 3, the coefficient of x in the first equation, ensures that x has the same coefficient in both equations:

$$6x - 10y = 6, \qquad 6x + 9y = 63$$

When we do the subtraction as before, the coefficient of x becomes $6 - 6$, so the term in x disappears, that of y becomes $-10 - 9 = -19$, and the right-hand side becomes $6 - 63 = -57$:

$$-19y = -57$$

so $y = -57/(-19) = 3$, and substituting this in

$$6x - 10y = 6$$

gives

$$6x - 30 = 6$$

and hence $x = (6 + 30)/6 = 36/6 = 6$.

An alternative way of achieving the same effect is to *divide* each equation by the coefficient of x (or whatever unknown is chosen for elimination) in that equation. In the above example, we would have:

$$x - 1.6667y = 1, \qquad x + 1.500y = 10.5$$

As this produces non-integer coefficients in the equations, it may seem rather ugly and unappealing; nonetheless, it is a good general approach. One reason for this is that in experimental problems (as opposed to invented ones in textbooks) the coefficients are hardly ever integers in the first place. Another is that division, rather than multiplication, tends to ensure that the calculation is done in terms of numbers not very different from 1, which decreases numerical inaccuracies, whether the calculation is done by hand or by computer. These numerical considerations may seem trivial, as indeed they usually are for simultaneous equations in only two unknowns, but they become of great importance in solving sets of equations in several unknowns.

6.4 **Determinants**

The methods I have described for solving simultaneous equations expressed in numerical terms can also be applied to equations in which the coefficients are not assigned numerical values. Consider the following general pair of linear simultaneous equations:

$$A_1 x + B_1 y + C_1 = 0$$

$$A_2 x + B_2 y + C_2 = 0$$

Multiplying the first by A_2 and the second by A_1 we have

$$A_1 A_2 x + A_2 B_1 y + A_2 C_1 = 0$$

$$A_1 A_2 x + A_1 B_2 y + A_1 C_2 = 0$$

and subtraction of the first equation from the second results in elimination of x:

$$(A_1 B_2 - A_2 B_1) y + (A_1 C_2 - A_2 C_1) = 0$$

Hence

$$-y = \frac{A_1 C_2 - A_2 C_1}{A_1 B_2 - A_2 B_1}$$

and, similarly,

$$x = \frac{B_1 C_2 - B_2 C_1}{A_1 B_2 - A_2 B_1}$$

A special notation has been devised to allow this general solution to a set of simultaneous equations to be expressed in a particularly tidy way, as follows:

$$\frac{x}{\begin{vmatrix} B_1 & C_1 \\ B_2 & C_2 \end{vmatrix}} = \frac{-y}{\begin{vmatrix} A_1 & C_1 \\ A_2 & C_2 \end{vmatrix}} = \frac{1}{\begin{vmatrix} A_1 & B_1 \\ A_2 & B_2 \end{vmatrix}}$$

The denominators of these expressions are called *determinants*. Any determinant is written as a square array of numbers or algebraic symbols between vertical lines and it has a numerical or algebraic value that can be found by applying a general rule. Although a determinant can be a square of any size, I shall only consider 2×2 determinants, which not only are particularly simple but are also the most important for our purposes: the value of such a determinant is the product of the

Box 6.2 **Solving simultaneous equations with determinants**

1. Write down the pair of equations to be solved.

$$2x+3y=5$$
$$y-5=4x$$

2. Arrange them systematically, so that the unknowns appear in the same sequence in every equation (with appropriate gaps or zeroes if there are any unknowns missing from particular equations) and the constant as the last term on the left-hand side of each equation. For each equation the right-hand side should then consist of zero.

$$2x+3y-5=0$$
$$4x-y+5=0$$

3. Write down the solution in determinant form as a series of expressions that are all equal to one another. Each unknown appears once as numerator, with a numerator of 1 for the last; each denominator is obtained by writing a determinant consisting of the coefficients in the equations but *omitting the column for the quantity that appears in the numerator*; the signs of the numerators alternate.

$$\frac{x}{\begin{vmatrix} 3 & -5 \\ -1 & 5 \end{vmatrix}} = \frac{-y}{\begin{vmatrix} 2 & -5 \\ 4 & 5 \end{vmatrix}} = \frac{1}{\begin{vmatrix} 2 & 3 \\ 4 & -1 \end{vmatrix}}$$

4. Each determinant is then evaluated as above.

$$\frac{x}{10} = \frac{-y}{30} = \frac{1}{-14}$$

5. Hence

$$\frac{x}{10} = \frac{1}{-14}, \quad x = -\frac{5}{7}$$

and

$$\frac{-y}{30} = \frac{1}{-14}, \quad y = \frac{30}{14} = \frac{15}{7}$$

top-left and bottom-right elements minus the product of the bottom-left and top-right:

$$\begin{vmatrix} a_1 & b_1 \\ a_2 & b_2 \end{vmatrix} \equiv \left(a_1 \quad \cdot \quad b_2 \right) - \left(\quad \cdot \quad b_1 \atop a_2 \right) = a_1 b_2 - a_2 b_1$$

To write the solution of a set of equations in determinant form we proceed in accordance with the set of rules listed in Box 6.2, which are illustrated by reference to the following pair of equations:

$$2x + 3y = 5$$
$$y - 5 = 4x$$

Once the determinants are known it is an easy matter to write down the individual solutions by multiplying the right-hand expression by each of the denominators in turn:

$$x = \frac{10}{-14} = -\frac{5}{7}$$

$$y = \frac{-30}{-14} = \frac{15}{7}$$

It often seems more natural to write each of the original set of equations with the constant on the right-hand side rather than as the last term on the left-hand side, i.e. (for the same example):

$$2x + 3y = 5$$
$$4x - y = -5$$

If this is done the solution can be written in the same way as before *except* that the sign of the constant is the opposite of what it would be if the equations were written in standard form. For two simultaneous equations this means that the last numerator is −1 rather than 1:

$$\frac{x}{\begin{vmatrix} 3 & 5 \\ -1 & -5 \end{vmatrix}} = \frac{-y}{\begin{vmatrix} 2 & 5 \\ 4 & -5 \end{vmatrix}} = \frac{-1}{\begin{vmatrix} 2 & 3 \\ 4 & -1 \end{vmatrix}}$$

The determinant in the right-hand (constant) expression has a special importance for any set of equations. It appears in the denominator of the solution for every unknown, and its value indicates whether a unique solution to the set of equations exists: if it is zero then dividing by it is a division by zero, so there is no unique solution and the equations are singular. It is, therefore, useful to be able to recognize characteristics of a determinant whose value is zero. The following rules apply to *all* determinants, not just those of order 2:

1   If any row or column consists entirely of zeroes, the determinant is zero. This is always true for a set of equations in which the number of equations is different from the number of unknowns.

2   If any two rows are identical the determinant is zero. This occurs if any equation is identical to any other, with the result that the number of *different* equations is less than the number of unknowns.

3   If any two columns are identical, the determinant is zero. This occurs if the information given about any two unknowns is identical, so that they cannot be distinguished.

4 If any row or column can be obtained by multiplying all of the elements in another row or column by the same constant, the determinant is zero. For rows, this corresponds to the case where one equation can be obtained from another merely by multiplying every term by the same constant.

5 If any row or column is a linear function of two or more other rows or columns, the determinant is zero. This means that a row (or column) can be obtained by writing down a sum of combinations of other rows (or columns). For example, in the determinant

$$\begin{vmatrix} 2 & 3 & 7 \\ 1 & 4 & 6 \\ 0 & 1 & 1 \end{vmatrix}$$

we can obtain any element in the right-hand column by adding the element in the middle column to twice the element in the left-hand column: $7 = 3 + 2 \times 2$; $6 = 4 + 2 \times 1$; $1 = 1 + 2 \times 0$.

It is immediately obvious that conditions 1 to 4 are simple enough to be able to tell by inspection if they are satisfied, but condition 5 is much more complicated. It is not possible to satisfy it at all with a determinant of order 2, i.e. for a pair of simultaneous equations, but for larger determinants it becomes very important. Indeed, in complicated cases it is much more likely that condition 5 will be satisfied than that any of the others will. We can, however, ignore it in a simple course of mathematics, remembering only that it means that a determinant may be zero and the corresponding equations insoluble even if none of the more obvious tests indicate a singular determinant.

6.5 Quadratic equations

With the equation

$$(x-5)(x-1) = 0$$

it is obvious that the left-hand side is equal to zero if *either* of the two terms in parenthesis is zero, i.e. $x - 5 = 0$ *or* $x - 1 = 0$, because anything multiplied by zero is zero: if one term in a product is zero it does not matter what the others are. Thus there are two values of x, $x = 5$ and $x = 1$, that satisfy the equation; these are often known as the *roots* of the equation. When the equation is written in this way the solution is obvious, but unfortunately equations of this kind usually do not appear initially as a set of separate linear factors, but with these factors multiplied out. When the left-hand side of the above equation is multiplied out it becomes a typical *quadratic equation*:

$$x^2 - 6x + 5 = 0$$

As this is exactly the same equation as the original one it must have the same pair of solutions, $x=5$ or $x=1$, but this is no longer obvious from inspection, so we need to examine ways of solving quadratic equations that are written in the multiplied-out form.

In principle, a quadratic equation can be solved by reversing the multiplication above, i.e. by finding a pair of factors that give the appropriate expression when multiplied together: as long as we can recognize that x^2-6x+5 is the result of multiplying out $(x-5)(x-1)$ then we can recognize the solution immediately. This is usually the method that is taught first at school and in artificially constructed examples it works well. Unfortunately, however, it is virtually useless in the real world because it is so rare for an equation encountered in a real problem to have *rational factors* that it is a waste of time looking for them. (A rational number is one that can be written as a ratio of two integers, like $-5/7$ in the last section.)

We shall not waste time therefore looking at the simple (but in practice useless) method of factorization for solving quadratic equations but will go straight to the general method that will always work. Let us first consider how we might set about solving the quadratic equation above if we did not know that there were simple linear factors:

$$x^2-6x+5=0$$

If we compare the left-hand side with the expression for $(x-3)^2$,

$$(x-3)^2=x^2-6x+9$$

we should notice that the first two terms are the same; only the constant is different. We can make all three terms identical by adding an appropriate constant, 4 in this example, and if we add it to the left-hand side we must, of course, add it to the right-hand side as well:

$$x^2-6x+9=4$$

i.e.

$$(x-3)^2=4$$

which can easily be solved by taking square roots of both sides (not forgetting that 4 is the square not only of 2 but also of -2):

$$x-3=\pm2$$

Hence $x=5$ or $x=1$, exactly as before. In this example, the right-hand side proved to be the square of an integer (because I chose numbers that would come out like that), but in general this would *not* be the case; for arbitrarily chosen numbers the result would be far more likely to give a right-hand side that was not a perfect square. For example, if we applied the same method to the equation

$$x^2-6x+4=0$$

the result would be

$$(x-3)^2 = 5$$

which has no solution in terms of rational numbers. Nonetheless, it is perfectly soluble as long as we can calculate a square root (a trivial problem for almost any modern calculator).

The method we have used, therefore, is a *general method* that will always lead to the solutions of any quadratic equation if solutions exist. It is called *completing the square* and it forms the basis of what is actually the most widely used method in scientific problems, which is *solution by formula*. This is just an application of the same ideas to a general equation expressed in terms of symbols instead of numbers:

$$ax^2 + bx + c = 0$$

The three constants a, b and c represent any numbers that we might want to write. Thus, this is a *generalization* of the original example and, if we can solve it, then we ought to be able to solve *any* quadratic equation.

First we note that in the numerical examples the first term was just x^2, i.e. the coefficient of x^2 was 1, but in general this will not be the case, and so we begin by making it the case, by dividing all terms by a, the original coefficient of x^2:

$$x^2 + \frac{bx}{a} + \frac{c}{a} = 0$$

In the numerical example, we noticed that the first two terms of the left-hand side of the equation were the same as the first two terms of the square of $(x-3)$. In this example the number -3 did not spring from nowhere but was actually one-half of -6, i.e. of the coefficient of x in the equation to be solved. In the general equation, we make a corresponding comparison with the square of $(x + b/2a)$:

$$\left(x + \frac{b}{2a}\right)^2 = x^2 + \frac{bx}{a} + \frac{b^2}{4a^2}$$

The first two terms are exactly the same as those on the left-hand side of the equation to be solved, but the third is different. As these two expressions are equal, we will get zero if we subtract one from the other, and we can then equate the difference to the original expression for zero:

$$x^2 + \frac{bx}{a} + \frac{c}{a} = x^2 + \frac{bx}{a} + \frac{b^2}{4a^2} - \left(x + \frac{b}{2a}\right)^2$$

Cancelling $x^2 + \frac{bx}{a}$, which occurs on both sides of the equation,

$$\frac{c}{a} = \frac{b^2}{4a^2} - \left(x + \frac{b}{2a}\right)^2$$

rearranging,

$$\left(x+\frac{b}{2a}\right)^2 = \frac{b^2}{4a^2} - \frac{c}{a}$$

and bringing the right-hand side over a common denominator,

$$\left(x+\frac{b}{2a}\right)^2 = \frac{b^2-4ac}{4a^2}$$

we have an equation that can easily be solved: the left-hand side is a perfect square, so we know its square root already, and the right-hand side is just a number (if we assign numerical values to a, b and c), so in principle we can obtain its square root with a calculator. Taking square roots of both sides, therefore, we have (again remembering that there are two different numbers that will give the right-hand side if we square them):

$$x+\frac{b}{2a} = \frac{\pm\sqrt{b^2-4ac}}{2a}$$

It is now just a simple matter of rearrangement to get the expressions for x:

$$x = \frac{-b \pm \sqrt{b^2-4ac}}{2a}$$

The word *expressions* is in the plural because this represents *two* solutions, as the sign $\pm$ can be interpreted as *either* plus *or* minus. It is necessary here because either of the two expressions with square equal to b^2-4ac will satisfy the equation. The square-root sign by itself does not convey this double meaning because it is defined to be unambiguous and specifies the positive square root only.

This result is the general solution for a quadratic equation. Whether it is worth memorizing depends on how often you plan to solve quadratic equations, but it is important to understand how to use it even if you do not memorize it.

The quantity b^2-4ac is known as the *discriminant* of the equation $ax^2+bx+c=0$. It is a useful quantity because its value tells us something in advance about the solutions of the equation (Table 6.1). If it is positive, the equation has two real and

Table 6.1 Use of the discriminant to explore the roots of a quadratic equation.

Value	Interpretation	Example	Discriminant	Roots
Positive	Two real roots	$x^2+3x-2=0$	$9+8=17$	$0.562, -3.562$
Negative	No real roots	$x^2+3x+4=0$	$9-16=-7$	None
Zero	Two equal roots	$x^2-6x+9=0$	$36-36=0$	$3, 3$
Perfect square	Two rational roots	$x^2-2x-3=0$	$4+12=16=4^2$	$-1, 3$

The *discriminant* of a quadratic equation $ax^2+bx+c=0$ is the quantity b^2-4ac, and its value allows information about the solutions to the equation to be deduced without carrying out the whole solution.

unequal solutions; if it is zero, there are two real solutions, but they are identical; if it is positive and a perfect square, such as 25, the solutions are rational and the simple factorization method mentioned at the beginning of this section could be used for solving the equation; if it is negative, there are no real solutions. This last result follows from the fact that a negative number does not have a real square root, and it is important because it shows that one can have an apparently elementary equation that has no solution. Just because we can write down a simple equation does not mean it has a solution.

For elementary purposes, we can omit the word *real* from the preceding statements, because in elementary work only real solutions are considered as solutions at all. In more advanced mathematics, it becomes convenient to define the so-called *imaginary* numbers as the square roots of negative real numbers, and *complex* numbers as the results of adding real and imaginary numbers together. However, although these kinds of numbers have important applications in some scientific contexts—in quantum mechanics and in the theory of electrical circuits, for example—it is difficult to think of an application in elementary biochemistry; so, I shall not consider them further in this book.

From the mathematical point of view both of the solutions to a quadratic equation are equally valid and there is no reason to say that either is any 'better' than the other. It is different in scientific contexts because one usually finds that only one of the solutions makes any physical sense, and this can usually be identified by inspection. Consider, for example, the equilibrium between the three adenylates:

$$2\,ADP = ATP + AMP$$

which has an equilibrium constant of about $\frac{1}{2}$ under physiological conditions, i.e.

$$\frac{[ATP][AMP]}{[ADP]^2} = \frac{1}{2}$$

Suppose now we mix $8\,mM$ ATP with $2\,mM$ ADP and add a little of the enzyme adenylate kinase to catalyses the conversion to equilibrium: what will be the final concentrations of the three adenylates? If we put $[AMP] = x$, the stoichiometry of the reaction requires that $[ATP] = 8 + x$, $[ADP] = 2 - 2x$, and so the equilibrium expression can be written as

$$\frac{(8+x)x}{(2-2x)^2} = \frac{1}{2}$$

and with some rearrangement ('cross-multiply', i.e. multiply the left-hand side by 2 and the right-hand side by the expression that appears as the denominator on the left-hand side; then apply the operations listed in Box 6.1) this becomes a conventional quadratic equation:

$$x^2 - 12x + 2 = 0$$

Applying the formula derived above,

$$x = \frac{12 \pm \sqrt{12^2 - 8}}{2} = 6 \pm 5.83 = 11.83 \ \textit{or} \ 0.169 \, \text{mM}$$

At first sight, both these solutions are physically meaningful because they are both concentrations that AMP could conceivably have. However, calculation of the corresponding ADP and ATP concentrations shows that

if $[AMP] = 11.8 \, \text{mM}$, then $[ADP] = -21.6 \, \text{mM}$, $[ATP] = 19.8 \, \text{mM}$
if $[AMP] = 0.169 \, \text{mM}$, then $[ADP] = 1.66 \, \text{mM}$, $[ATP] = 8.17 \, \text{mM}$

As concentrations in the real world cannot be negative, only the second solution is physically acceptable and the first must be discarded. As a general rule, one should always check that the solution of an equation makes physical as well as mathematical sense. It is perfectly possible to have results that are entirely sound mathematically but still make no physical sense.

6.6 **Graphical solution of equations**

Sometimes we have to solve equations that are more complex than quadratics. Exact analytical solutions for certain kinds of higher-order equations exist, but these are hardly ever used. The practical choice is between *graphical solution* and *Newton's method*, which I shall discuss in the next two sections.

Let us consider the following equation as an example of the sort of equation that would be very difficult (if not impossible) to solve analytically:

$$x^2 - 3x + \ln x = -1$$

However, we can solve it quite easily by a graphical method. We begin by rewriting it so that the right-hand side is zero:

$$x^2 - 3x + \ln x + 1 = 0$$

if we define a function $f(x)$ as the left-hand side:

$$f(x) = x^2 - 3x + \ln x + 1$$

This function will have different values at different values of x, of course, but it will have the value zero when x is a solution to the original equation. If we plot $f(x)$ against x (Fig. 6.1), the line crosses the x-axis at about $x = 2.2$. As the x-axis is an expression of the relationship $f(x) = 0$, it follows that $x = 2.2$ is an approximate solution to the equation.

The advantage of graphical solution is that it is easy to understand and use, but it has compensating disadvantages also. First, it is laborious, as the function has to be evaluated at numerous different x values, many of which turn out not to be close to the solution. Second, it is limited by the accuracy of one's draughtsmanship and

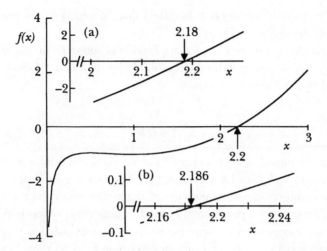

Fig. 6.1 Graphical solution to the equation $x^2 - 3x + \ln x + 1 = 0$. Plotting the function $f(x) = x^2 - 3x + \ln x + 1$ against x allows the solution to be estimated from the point where the curve crosses the x-axis. (**a**) The accuracy of the solution can be increased by replotting on a larger scale in a range of x values close to the estimated solution, and (**b**) this can be repeated as many times as necessary to arrive at any desired accuracy.

ability to read off the solution accurately. One can improve the accuracy of the solution by drawing another graph on a larger scale covering only a narrow range of x values, say from 2.0 to 2.3, as shown in inset (a) and, if one takes the trouble, there is no limit to the accuracy that can eventually be attained. This adds substantially to the labour required, however, and it is normally better to proceed to Newton's method (next section) if one needs a more accurate solution than the one given by the preliminary graph.

One can also use graphical methods for solving simultaneous equations. Although this is not a particularly convenient way of dealing with straightforward linear simultaneous equations of the sort discussed earlier in this chapter, it is still important, for two reasons. First, it provides one of the simplest ways of solving non-linear simultaneous equations. Second, the method illustrates the principle underlying certain graphical techniques used in enzyme kinetics, most notably the Dixon plot for determining inhibition constants. For this reason, I shall discuss the method briefly. Consider the equations

$$5x + 7y = 6$$
$$2x - y = 1$$

Each of these defines a straight line if y is plotted against x. Every point on the first line has coordinates x and y that specify an (x, y) pair that satisfies the first equation, and every point on the second line has coordinates x and y that specify an (x, y) pair that satisfies the second equation. There is only one point that lies on both lines, the point of intersection, and the values of x and y at this point are those that satisfy both equations (Fig. 6.2).

The Dixon plot is a more complex application of the same principle. *Competitive inhibition* is the name given to the kind of inhibition of an enzyme-catalysed reaction defined by the following equation:

$$v = \frac{Va}{K_m(1 + i/K_i) + a}$$

in which v is the rate observed at concentrations i and a of inhibitor and substrate respectively, and V, K_m, and K_i are constants. The Dixon plot is especially convenient in the common experimental circumstance where our main objective is to measure K_i, the other two parameters V and K_m being either of less immediate concern or known independently. If we have values of v measured at several i values at each of two a values, a_1 and a_2, we can determine K_i by plotting $1/v$ against i at both a values. If we use v_1 and v_2 as symbols for the v values measured at $a = a_1$ and $a = a_2$ respectively, the two straight lines are given by

$$\frac{1}{v_1} = \frac{K_m}{Va_1} + \frac{1}{V} + \frac{K_m}{VK_i a_1} \cdot i$$

$$\frac{1}{v_2} = \frac{K_m}{Va_2} + \frac{1}{V} + \frac{K_m}{VK_i a_2} \cdot i$$

The point of intersection of the two lines defines the value of i at which $1/v_1 = 1/v_2$ and hence the value at which the two right-hand sides are equal:

$$\frac{K_m}{Va_1} + \frac{1}{V} + \frac{K_m}{VK_i a_1} \cdot i = \frac{K_m}{Va_2} + \frac{1}{V} + \frac{K_m}{VK_i a_2} \cdot i$$

Despite its messy appearance, this is just a simple linear equation in i which may be solved in the ordinary way: first we subtract $1/V$ from both sides to remove the middle terms, and then we move terms in i to the left-hand side and terms without i to the right-hand side:

$$\frac{K_m}{VK_i a_1} \cdot i - \frac{K_m}{VK_i a_2} \cdot i = \frac{K_m}{Va_2} - \frac{K_m}{Va_1}$$

and factorizing, we find:

$$\frac{K_m}{V}\left(\frac{1}{a_1} - \frac{1}{a_2}\right) \cdot \frac{i}{K_i} = -\frac{K_m}{V}\left(\frac{1}{a_1} - \frac{1}{a_2}\right)$$

Hence $i/K_i = -1$, or $i = -K_i$. It follows that the value of i at the point of intersection of the two lines gives $-K_i$ (Fig. 6.3).

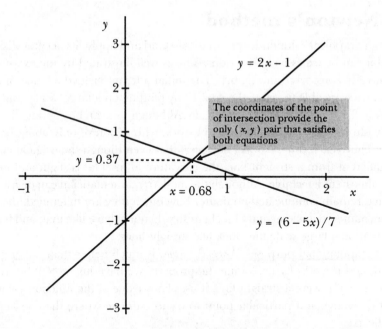

Fig. 6.2 Graphical solution of the simultaneous equations $5x+7y=6$, $2x-y=1$. The two straight lines represent $y=(6-5x)/7$ and $y=2x-1$, and their point of intersection, the only point that lies on both lines, provides the coordinates of the only (x, y) pair that satisfies both equations, i.e. $x=0.68$, $y=0.37$.

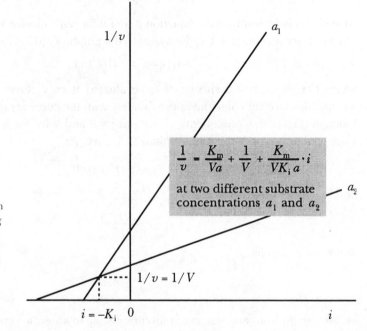

Fig. 6.3 The Dixon plot for determining a competitive inhibition constant. This is a specifically biochemical application of the method illustrated in Fig. 6.2.

6.7 **Newton's method**

Although graphical solution is easy to understand and apply, its accuracy is limited and it can be excessively laborious. This is well illustrated by the example we have already considered in Fig. 6.1. The initial attempt yielded a value of about 2.2, but we were able to refine this to 2.18 by plotting on a larger scale, and additionally to 2.186 with an even larger scale. Although this can be continued as long as we wish, it clearly represents a lot of work, with a gain of only about one significant figure each time. Notice, however, that even though the original curve is quite different from a straight line, the curvature in inset (a) is slight and in inset (b) it is almost undetectable. Although the modern mathematician can devise functions that remain complicated no matter how much they are magnified, the functions encountered in elementary biochemistry do not behave like that, and it is safe to say that on a large scale they look like straight lines.

For computational purposes, *Newton's method* is much better than the graphical method, and it works by exploiting the property we have just noted: by treating any curve as if it were a straight line, it uses knowledge of the function value and its first derivative at a particular point to try to estimate where the line ought to cross the axis.

I shall use as an example the same equation as we used for studying the graphical method:

$$x^2 - 3x - \ln x = -1$$

Then if we define a function $f(x)$ as

$$f(x) = x^2 - 3x + \ln x + 1$$

we have, as before, solved the equation if we find a value of x for which $f(x) = 0$. At any arbitrary starting point x_0 the *tangent* to the graph of $f(x)$ against x is given by

$$y = f(x_0) + (x - x_0)f'(x_0)$$

where $f'(x_0)$ is the first derivative of $f(x)$ evaluated at $x = x_0$. Over a finite range of x values this tangent coincides approximately with the curve representing the true function (Fig. 6.4). Consequently, if we put $y = 0$ and solve for x we should get a value x_1 that is nearer the solution than x_0, i.e. we put

$$f(x_0) + (x_1 - x_0)f'(x_0) = 0$$

and rearrange to give

$$x_1 = x_0 - \frac{f(x_0)}{f'(x_0)}$$

and more generally,

$$x_{n+1} = x_n - \frac{f(x_n)}{f'(x_n)}$$

i.e. we can go from any starting point (not just x_0) to a new approximation.

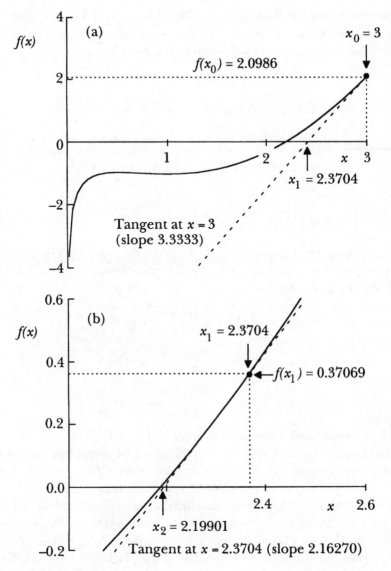

Fig. 6.4 Graphical illustration of Newton's method for solving the equation $x^2 - 3x + \ln x + 1 = 0$. This is the same problem as the one illustrated in Fig. 6.1, but here the graph is not actually used to find the solution; it just illustrates what happens during the calculation. (a) An arbitrary guess $x_0 = 3$ is taken as a starting point, and the curve is assumed to resemble the tangent at this x value, and the point where the tangent crosses the x-axis provides an improved guess, $x_1 = 2.3704$. (b) The calculation is repeated used the improved guess as starting point. It can be repeated as many times as necessary to produce any desired accuracy.

This is the fundamental relationship of Newton's method. Let us see how it can be used for solving the equation that we set out with. For that equation, differentiation of $f(x)$ with respect to x shows that $f'(x)$ is given by

$$f'(x) = 2x - 3 + \frac{1}{x}$$

If we choose an arbitrary starting point of $x_0 = 3$, then

$$f(x_0) = f(3) = 3^2 - 3 \cdot 3 + \ln 3 + 1 = 9 - 9 + 1.098\,6 + 1 = 2.098\,6$$

$$f'(x_0) = 2 \cdot 3 - 3 + \frac{1}{3} = 6 - 3 + 0.333\,3 = 3.333\,3$$

$$x_1 = x_0 - \frac{f(x_0)}{f'(x_0)} = 3 - \frac{2.098\,6}{3.333\,3} = 2.370\,4$$

We can see from Fig. 6.4 that although this is not the correct solution it is closer to it than $x_0 = 3$. So we proceed with a second approximation, replacing x_0 with x_1 and x_1 with x_2, and so on:

$$x_2 = 2.370\,4 - \frac{0.370\,69}{2.162\,70} = 2.199\,01$$

$$x_3 = 2.199\,01 - \frac{0.026\,63}{1.852\,78} = 2.184\,64$$

$$x_4 = 2.184\,64 - \frac{0.000\,19}{1.827\,02} = 2.184\,54$$

One can continue until any desired accuracy is achieved.

In this example one can recognize that the successive approximations are nearer and nearer to the solution by the fact that the values of $f(x)$ in the numerator of the second term get smaller and smaller with each approximation: 2.098 6, 0.370 69, 0.026 63, 0.000 19. Sometimes this does not happen and instead of getting better and better the successive approximations given by Newton's method get worse and worse. This may happen, for example, if $f'(x_0)$ is close to zero, so that x_0 is close to a maximum or a minimum of $f(x)$. Figures 6.1 and 6.4 illustrate how easily this can happen: as the value of $f(x)$ changes very little for x values in the range 0.4–1.2 the curve is very flat in this region, i.e. $f'(x_0)$ is close to zero; any tangent drawn in this region will be almost parallel to the x-axis and so following it until the x-axis is reached implies a very large change in x.

One can usually cure this problem when it occurs by trying a different value of x_0. In severe cases a preliminary graphical exploration as described in the preceding section should reveal a suitable range of values to try. For equations with two or more solutions, Newton's method does not guarantee to find all of them; it will usually (though by no means necessarily) proceed towards the solution closest to x_0. Again, graphical exploration is useful in such cases to ensure that no solutions have been overlooked.

6.8 **Approximate methods**

Textbook illustrations of methods for solving equations commonly present examples in which the coefficients are numbers in the range 1 to 10, and often they are integers. Real life is often very different and one may, instead, have to solve equations with coefficients ranging over several orders of magnitude. This has two consequences: first, methods that work well with textbook examples may work badly in real life; and second, there is much more scope for the use of *approximations* than one might expect from the study of elementary examples.

To illustrate these points, let us consider how to calculate the pH of a 0.1 M solution of monosodium glutamate, given that glutamic acid has three ionizations, with pK_a values of 2.3, 4.3 and 9.7. These convert to K_a values of 5×10^{-3}, 5×10^{-5}, and 2×10^{-10}, respectively, and so the equilibria can be formulated as follows:

$$G^{1+} \underset{\xrightarrow{\hspace{1cm}}}{\xrightarrow{5 \times 10^{-3}}} G^0 \underset{\xrightarrow{\hspace{1cm}}}{\xrightarrow{5 \times 10^{-5}}} G^{1-} \underset{\xrightarrow{\hspace{1cm}}}{\xrightarrow{2 \times 10^{-10}}} G^{2-}$$

where G^{1+} represents protonated glutamic acid with a net charge of $+1$, G^0 represents glutamic acid with a net charge of zero, etc. (actually three different states with charge zero can be drawn, but for calculating pH values it is not necessary to distinguish between them, and G^0 comprises all of them; similarly, G^{1-} comprises three states with charge -1). From the definitions of the equilibrium constants we can write

$$[G^0] = 5 \times 10^{-3} \times \frac{[G^{1+}]}{[H^+]}$$

$$[G^{1-}] = 5 \times 10^{-5} \times \frac{[G^0]}{[H^+]} = 2.5 \times 10^{-7} \times \frac{[G^{1+}]}{[H^+]^2}$$

$$[G^{2-}] = 5 \times 10^{-10} \times \frac{[G^{1-}]}{[H^+]} = 5 \times 10^{-17} \times \frac{[G^{1+}]}{[H^+]^3}$$

Notice that we have five unknowns here (the concentrations of H, G^{1-}, G^0, G^{1+}, and G^{2+}), but only three equations: clearly, we cannot solve them without additional information. This is supplied by two stoichiometric requirements that must be satisfied. First, the total concentration of the different forms of glutamic acid is a constant:

$$[G^{1+}] + [G^0] + [G^{1-}] + [G^{2-}] = 0.1 \, M$$

and second, the total concentration of negative charge must equal the concentration of Na^+ ions, which is also 0.1 M:

$$-[G^{1+}] + [G^{1-}] + 2[G^{2-}] = 0.1 \, M$$

We now have five simultaneous equations in the five unknowns, $[G^{1+}]$, $[G^0]$, $[G^{1-}]$, $[G^{2-}]$ and $[H^+]$. At first sight this is a much more complicated example than any considered before, but as the first three equations express $[G^0]$, $[G^{1-}]$, and $[G^{2-}]$ in terms of $[G^{1+}]$ it is easy to eliminate these three unknowns by substituting them in the last two equations:

$$[G^{1+}]\left(1+\frac{5\times10^{-3}}{[H^+]}+\frac{2.5\times10^{-7}}{[H^+]^2}+\frac{5\times10^{-17}}{[H^+]^3}\right)=0.1\,\text{M}$$

$$[G^{1+}]\left(-1+\frac{2.5\times10^{-7}}{[H^+]^2}+\frac{1\times10^{-16}}{[H^+]^3}\right)=0.1\,\text{M}$$

$[G^{1+}]$ can now be eliminated by dividing one equation by the other, though before doing this we should check that we are not dividing by zero: dividing both sides of an equation by the same quantity (or by quantities that are equal to one another, as in this example) is a legitimate step *only* if the divisor cannot be equal to zero. In this case the right-hand side is 0.1 M, which is not zero, and as the left-hand side is equal to it, it cannot be zero either, so the division is legitimate:

$$\frac{1+\dfrac{5\times10^{-3}}{[H^+]}+\dfrac{2.5\times10^{-7}}{[H^+]^2}+\dfrac{5\times10^{-17}}{[H^+]^3}}{-1+\dfrac{2.5\times10^{-7}}{[H^+]^2}+\dfrac{1\times10^{-16}}{[H^+]^3}}=1$$

We now multiply both sides of the equation by the denominator of the expression on the left-hand side:

$$1+\frac{5\times10^{-3}}{[H^+]}+\frac{2.5\times10^{-7}}{[H^+]^2}+\frac{5\times10^{-17}}{[H^+]^3}=-1+\frac{2.5\times10^{-7}}{[H^+]^2}+\frac{1\times10^{-16}}{[H^+]^3}$$

and remove the remaining fractions by multiplying all terms by $[H^+]^3$:

$$[H^+]^3+5\times10^{-3}[H^+]^2+2.5\times10^{-7}[H^+]+5\times10^{-17}$$
$$=-[H^+]^3+2.5\times10^{-7}[H^+]+1\times10^{-16}$$

Finally, gathering like terms together gives a *cubic equation* (i.e. an equation in which the highest-order term is the third power, or *cube*, of the unknown) in a single unknown, $[H^+]$:

$$2[H^+]^3+5\times10^{-3}[H^+]^2-5\times10^{-17}=0$$

On the face of it, this is a straightforward cubic equation and we might expect to be able to solve it by Newton's method. However, the enormous range of coefficients from 2 to 5×10^{-17} makes it a very different proposition from the equations we have considered earlier in this chapter. For this example, the function value and its first derivative are as follows:

$$f([H^+]) = 2[H^+]^3 + 5 \times 10^{-3}[H^+]^2 - 5 \times 10^{-17}$$

$$f'([H^+]) = 6[H^+]^2 + 10^{-2}[H^+]^2$$

Let us choose $[H^+]_0 = 1$ M as a starting guess. As we shall see, this is actually a very poor starting guess, but if we treat it as a purely mathematical exercise, without using any knowledge of fundamental chemistry that might allow a better guess, it is not very likely that we shall start at a point close to the right answer. This then gives

$$f([H^+]_0) = 2 + 5 \times 10^{-3} - 5 \times 10^{-17} = 2.005$$

(ignoring 5×10^{-17} at this stage as it is obviously negligible compared with the other two terms), and

$$f'([H^+]_0) = 6 + 10^{-2} = 6.01$$

So

$$[H^+]_1 = [H^+]_0 - \frac{f([H^+]_0)}{f'([H^+]_0)} = 1 - \frac{2.005}{6.01} = 6.66 \times 10^{-1} \, \text{M}$$

$$[H^+]_2 = [H^+]_1 - \frac{f([H^+]_1)}{f'([H^+]_1)} = 0.666 - \frac{0.594}{2.671} = 4.44 \times 10^{-1} \, \text{M}$$

$$[H^+]_3 = [H^+]_2 - \frac{f([H^+]_2)}{f'([H^+]_2)} = 0.444 - \frac{0.176}{1.187} = 2.96 \times 10^{-1} \, \text{M}$$

and so on. Progress is in the right direction but is painfully slow. In fact, it is not until the 30th approximation that we have even one significant figure correct:

$$[H^+]_{30} = [H^+]_{29} - \frac{f([H^+]_{29})}{f'([H^+]_{29})}$$

$$= 1.793 \times 10^{-7} - \frac{1.107 \times 10^{-16}}{1.793 \times 10^{-9}} = 1.175 \times 10^{-7} \, \text{M}$$

$$[H^+]_{31} = 1.013 \times 10^{-7} \, \text{M}$$

$$[H^+]_{32} = 1.000 \times 10^{-7} \, \text{M}$$

All subsequent approximations are very close to this last, showing that the correct concentration is 1.000×10^{-7} M, i.e. the pH is 7.0.

We can see by inspection that the answer is equal to the mean of the two higher pK_a values, i.e. $7.0 = (4.3 + 9.7)/2$, and we might well enquire whether this is just a coincidence, or whether we could have arrived at it by a quicker and easier route. Indeed, we can: the same enormous range of coefficients in the original equation that caused Newton's method to perform so badly allows us to introduce approximations that are virtually exact in this sort of problem even though the corresponding ones would be quite improper in an elementary example. To decide the best approximation, it is best to use our knowledge of chemistry. Because of the 0.1 M concentration of Na^+ ions, we know that the average charge on the glutamic acid species must be -1, so it is reasonable to guess that the predominant species will be G^{1-} and that G^{1+} will be present at negligible concentration.

If we ignore G^{1+} then we have a set of four simultaneous equations in four unknowns, and if we eliminate all except $[H^+]$ in the same way as before the resulting equation,

$$[H^+]^2 - 10^{-14} = 0$$

yields the solution $[H^+] = 10^{-7}$ M, or pH $= 7.00$, immediately.

The contrast between these two ways of solving the same problem could hardly be more striking. By applying a simple approximation at the beginning, we were able to obtain a solution that is so nearly exact that it is scarcely an approximation at all. Virtually all pH problems can be dealt with in this way with little difficulty. Even if several ionizations are involved, there are usually no more than three ionic states that need to be considered, and these can be identified by inspection.

Nonetheless, we may sometimes be faced with problems that are mathematically similar to the glutamic acid example but in which we have no chemical intuition to suggest what the answer is likely to be. In such cases, are there any purely mathematical considerations that can guide us to a better starting guess? To answer this, we should note that in a polynomial expression the higher-order terms usually vary much more steeply with the unknown variable than the lower-order terms do. We ought therefore to be able to deduce a rough value for the solution by ignoring all terms except the constant and the highest-order unknown term. In the case of the glutamic acid example, this would give just:

$$2[H^+]^3 - 5 \times 10^{-17} \approx 0$$

or

$$[H^+]^3 \approx 2.5 \times 10^{-17} \, M^3$$

from which we can deduce a starting guess of $[H^+]_0 = 2.924 \times 10^{-6}$ M, which, while far from perfect, is very much better than the one we tried first.

6.9 **Problems**

6.1 Rearrange the following equations to express x in terms of y:

(a) $y = x + 3$,   (b) $y = 2x^2 + 5$,
(c) $y = \ln 5x$,   (d) $y = 8 \exp(3x)$.

(e) $y = \dfrac{x+4}{x-1}$,   (f) $xy + 4 = 3x$

(g) $y = x^2 + 2x - 4$,   (h) $\dfrac{y}{x+1} = \dfrac{2}{x-1}$

(i) $\dfrac{y}{3x+1} - \dfrac{1}{x+1} = 1$

6.2 A metabolite B is produced in one reaction at a constant rate V_1 and is consumed in a second reaction at a rate $V_2[B]/(K_{m2} + [B])$. Assuming that a steady state is achieved in which [B] does not change, obtain an expression for [B]. How large would v_1 have to be for it to become impossible to establish a steady state?

6.3 Identify any of the following sets of simultaneous equations that are singular and solve the remainder:

(a) $2x + y = -1$, $x + y = -2$
(b) $x + 6y = 3.5$, $12y + 2x = 7$
(c) $8.32x + 1.03y + 39.85 = 0$, $7.18y - 2.22x = 96.14$
(d) $2.34x + 0.63y = 4.31$, $5.03x + 1.35y = 9.27$
(e) $px + qy = r$, $Px + Qy = R$
(f) $n\hat{a} + \hat{b}\ \Sigma x = \Sigma y$, $\hat{a}\ \Sigma x + \hat{b}\ \Sigma x^2 = \Sigma xy$

(Note: in problem (f) treat $\hat{a}$ and $\hat{b}$ as the unknowns and n, Σx, Σx^2, Σy, and Σxy as known quantities.)

6.4 (a) Solve the pair of equations in problem (6.3f) after substituting $x_1 = 1$, $y_1 = 1.27$; $x_2 = 2, y_2 = 2.47$; $x_3 = 3, y_3 = 3.62$; $x_4 = 4, y_4 = 5.08$; $n = 4$. Interpret Σx as the sum of x_i for $i = 1$ to n, and the other summations similarly.

(b) What would be the result of the calculation if we put $n = 1$ and used only x_1 and y_1?

6.5 Evaluate the following determinants:

(a) $\begin{vmatrix} 1 & 2 \\ 3 & 4 \end{vmatrix}$,   (b) $\begin{vmatrix} 4.71 & 1.28 \\ 6.43 & 5.11 \end{vmatrix}$,   (c) $\begin{vmatrix} -1 & -2 \\ 3 & -7 \end{vmatrix}$

(d) $\begin{vmatrix} 4 & 8 \\ 1 & 2 \end{vmatrix}$,   (e) $\begin{vmatrix} 0 & 0 & 0 \\ 1 & 3 & 6 \\ 4 & 2 & 1 \end{vmatrix}$,   (f) $\begin{vmatrix} 2.31 & 1.18 \\ -2.22 & 0.47 \end{vmatrix}$

(g) $\begin{vmatrix} 1.7 & 6.4 & 3.1 \\ 2.3 & 8.1 & 1.2 \\ 1.7 & 6.4 & 3.1 \end{vmatrix}$

6.6 Without carrying out a complete solution, calculate the discriminant of each of the following equations and determine whether there are (1) two unequal real roots, (1a) two unequal rational roots, (2) two equal roots, or (3) no real roots:

(a) $x^2 + 3x + 5 = 0$,   (b) $x^2 = 4x - 7$

(c) $2 + 3x = 4x^2$,   (d) $x^2 = 2x + 12$

(e) $3x(x + 4) = x + 6$,   (f) $\dfrac{2}{x+1} + \dfrac{3}{x+2} = 7$

6.7 The equilibrium constant for the reaction catalysed by hexokinase has the following value at pH 6.5:

$$\frac{[\text{glucose 6-phosphate}][\text{ADP}]}{[\text{glucose}][\text{ATP}]} = 230$$

If 5 mM-glucose is mixed with 4.5 mM-ATP and allowed to equilibrate at pH 6.5 in the presence of hexokinase, what is the final concentration of ADP?

6.8 A mixture of two biochemicals B and C has an absorbance in a 1 cm cuvette of 0.63 at 460 nm and of 0.52 at 500 nm. The molar absorbances, in $M^{-1}cm^{-1}$, are 1.03×10^3 for B, 4.57×10^3 for C at 460 nm; and 7.12×10^3 for B and 1.43×10^3 for C at 500 nm. Given that the absorbances of mixtures are additive, and that the absorbance of a solution with molar absorbance A, concentration c and pathlength d is Acd, calculate the concentrations of B and C.

6.9 Use a graphical method to obtain three approximate solutions to the equation $x^2 + 7x = 4/(3x + 2)$. Then use Newton's method to refine these to values correct to three places of decimals.

6.10 Estimate the pH of a 0.1 M solution of ammonium lactate, assuming that the pK_a of lactic acid is 3.86, the pK_a of the ammonium ion is 9.26, and $pK_w = 14.0$.

6.11 Estimate the pH of the solution obtained by mixing 3 mL of 0.1 M citric acid with 7 mL of 0.1 M NaOH. For citric acid, $pK_1 = 3.08$, $pK_2 = 4.74$, $pK_3 = 5.40$.

6.12 For quadratic equations that do not have simple solutions it is sometimes convenient to consider the sum and product of the two roots instead, which are usually simpler.

(a) Show this by deriving expressions for the sum and product of the roots of the general quadratic equation $ax^2 + bx + c = 0$.

(b) Hence write down the sum and product of the roots of the equation $x^2 - 7x + 5 = 0$.

CHAPTER 7

Partial differentiation

7.1 **The meaning of a partial derivative**

In science we often have to deal with two or more variables that have no necessary dependence on one another. For example, although Boyle's law defines the following relationship between the pressure p, volume V and temperature T of a mole of a perfect gas:

$$pV = RT$$

(where $R = 8.31\,\mathrm{J\,mol^{-1}\,K^{-1}}$ is the gas constant), it does not prevent us from varying any two of the three variables independently. We can have whatever pressure we like at whatever temperature we like, provided we accept whatever volume results, etc. If we are interested in how V varies with T, we can move p to the right-hand side of the equation:

$$V = \frac{RT}{p}$$

In this sort of circumstance, we cannot define an ordinary derivative, such as $\frac{\mathrm{d}V}{\mathrm{d}T}$, because this has no definite value unless we specify how p is to change with T. For example, if we decided that p was to be a constant independent of T then the right-hand side would be just a constant (R/p) multiplied by T, and so

$$\frac{\mathrm{d}V}{\mathrm{d}T} = \frac{R}{p}$$

but if we set up the system so that p increased in proportion to T with a constant coefficient a, then $p = aT$ and $V = R/a$, a constant, so

$$\frac{\mathrm{d}V}{\mathrm{d}T} = 0$$

More complicated relationships between p and T would produce different expressions for the derivative. This is clearly unsatisfactory, but we can overcome the difficulty by introducing the new concept of *partial differentiation*. The *partial derivative* of one variable with respect to another is defined as the result of differentiating the expression while treating all other *independent* variables as constants. So, for Boyle's law we would have:

$$\left(\frac{\partial V}{\partial T}\right)_p = \frac{R}{T_p} = \frac{V}{T}$$

$$\left(\frac{\partial V}{\partial p}\right)_T = -\frac{RT}{p^2} = -\frac{V}{p}$$

$$\left(\frac{\partial p}{\partial T}\right)_V = \frac{R}{V} = \frac{p}{T}$$

There are two points of notation to note here. First, as these are not ordinary derivatives, we cannot use the ordinary symbols $\frac{dV}{dT}$, etc., for them. Instead of the ordinary d, we use the so-called 'curly d', written as ∂, for partial differentiation. (This is not a Greek delta, incidentally, which would be δ and is used in mathematics to represent a small but finite increment, as in Chapter 4.) Second, we can indicate what variables are being held constant during the partial differentiation by showing them as subscripts, as in the expressions above. However, these are often obvious and may be omitted when no doubt is likely. In thermodynamics, they are frequently included as an aid to clarity. We call an expression such as $\frac{dV}{dT}$ 'partial dV by partial dT', although the second 'partial' is often omitted, so might say 'partial dV by dT'.

7.2 **Exact and inexact differentials**

In the previous section we saw that we cannot differentiate one variable with respect to another if the relationship between them is incompletely specified. The same applies to the reverse process, integration, and it is convenient to illustrate it with the same example of a perfect gas because of its great importance in thermodynamics.

Recalling that a *pressure* is a *force* divided by an *area* (or *length* squared), and *work* is the product of a *force* and a *length*, we can easily see that the work dW done (against the atmosphere, or a piston, or whatever) by a mole of perfect gas at pressure p expanding by an infinitesimal increment dV must be

$$dW = p\,dV$$

We might suppose, therefore, that we could calculate the work W done in a finite expansion from $V = V_1$ to $V = V_2$ as

$$W = \int_{V_1}^{V_2} p \, dV$$

However, just as in the previous section we were unable to evaluate $\dfrac{dV}{dT}$ without knowing how V depended on T, here we cannot evaluate $p \, dV$ without knowing how p depends on V. Further, because Boyle's law contains T as well as p and V, we must specify something about T before we can properly express p in terms of V. In this example, $p \, dV$ is an *inexact differential* and W can in fact have any value.

To integrate $p \, dV$, we must first define the system fully. Suppose we decide to consider an *isothermal* expansion, which is a physicist's way of saying that T is a constant by definition: if p is then written as RT/V, it is just a constant divided by V, and

$$W = \int_{V_1}^{V_2} p \, dV = RT \int_{V_1}^{V_2} \frac{dV}{V} = RT \, [\ln V]_{V_1}^{V_2} = RT \ln \left(\frac{V_2}{V_1} \right)$$

It is important to realize that we were able to do this integration, with T outside the integration sign, only by specifying T as a constant: if T varied during the expansion, the integration would be invalid.

Sometimes an apparently similar procedure leads to a definite result that does not depend on the way in which the particular change is carried out. For example, suppose we had set out to integrate not $p \, dV$ but $\left(\dfrac{p}{T} \right) dV$. Although at first sight this seems to be just as ill-defined as $p \, dV$, in fact it is a function of V only, because $p/T = R/V$ i.e.

$$\frac{p \, dV}{T} = \frac{R \, dV}{V}$$

The factor $1/T$, which converts the inexact differential $p \, dV$ into the *exact differential* $(p/T) \, dV$, is called an *integrating factor* for the inexact differential $p \, dV$.

Sometimes we find that although two differentials may be inexact, a simple function of them, such as their sum or difference, may be exact. For example, if no relationship between u and v is specified both $u \, dv$ and $v \, du$ are inexact differentials that cannot be integrated. Their sum, however, is $u \, dv$ and $v \, du$, which can immediately be recognized as $d(uv)$, an exact differential, if we recall the expression for the derivative of a ratio (Section 4.6).

This kind of relationship is of the greatest importance in thermodynamics. It is possible to convert any system (not just a sample of perfect gas) from one state to another by adding an indefinite amount of heat $\int dQ$ and by causing it to do an indefinite amount of work $\int dW$. These amounts of heat and work are indefinite because knowing the initial and final states does not allow us to calculate them: their values depend on the path taken between the two states. Nonetheless, it is a fact of observation that although dQ and dW are inexact differentials, their *difference* $dQ - dW$ is an exact differential. What this means is that although we may observe all kinds of values for the total heat $\int dQ$ and the total work $\int dW$ on passing from state 1 to state 2, we always observe exactly the same difference between them. This is clearly an important observation and is an expression of *the first law of thermodynamics*. The observation that $dQ - dW$ is an exact differential provides the basis for defining the *energy* U of a system, i.e. we define

$$dU = dQ - dW$$

It is important to realize that the fact that $dQ - dW$ is an exact differential is not something that could be demonstrated mathematically; it does not follow from any mathematical proof. It is a fact about the world that has been deduced by *observation*, i.e. it is an experimental result, not a mathematical one.

7.3 Least-squares fitting of the Michaelis–Menten equation

Although thermodynamics provides the main contexts in which students of chemistry and biochemistry encounter partial differentiation, it is a rather abstract subject that everyone finds difficult. It is useful, therefore, to consider a quite different example that has no obvious relationship to thermodynamics but still has a clear relevance to the practice of biochemistry.

Suppose we have a set of observations (a_i, v_i), for $i = 1, 2, 3, \ldots, n$, that fit the Michaelis–Menten equation apart from experimental error, i.e. suppose we can write

$$v_i = \frac{V \cdot a_i (1 + e_i)}{K_m + a_i}$$

where V and K_m are (unknown) constants and $(1 + e_i)$ is a factor representing the effect of experimental error. (We could also have included experimental error by *adding* an error term to the basic expression rather than by multiplying by an error factor. As my purpose in using this illustration has to do with mathematics rather than biochemistry, it is sufficient to say that we make this choice because it leads to simpler mathematics. As it happens, however, there are also good biochemical reasons for writing it this way.) In practice, we would not know V and K_m and would therefore wish to *estimate* them from the observed values of a_i and v_i; we might prefer to *measure* them instead, but the unknown magnitudes of the error terms prevent this.

By rearranging the above equation, we can readily express each e_i as follows:

$$e_i = \frac{K_m v_i + v_i a_i - V a_i}{V a_i}$$

$$= \frac{K_m v_i}{V a_i} + \frac{v_i}{V} - 1$$

We could proceed with the analysis with the same symbols if we wished, and the final results would be the same, but it is easier to follow if we introduce new variables $A = K_m / V$, $B = 1/V$, $x_i = v_i / a_i$:

$$e_i = Ax_i + Bv_i - 1$$

We can define the 'best' values of A and B as those that make some suitable measure of the e_i as small as possible. First, however, we must choose a function to measure this. It is not sufficient simply to add all the e_i's together, because some are positive and some are negative, so their sum can be small or zero without the individual e_i's having to be small. So, we commonly take the *sum of squares S*, defined as

$$S = \sum_{i=1}^{n} e_i^2 = \sum_{i=1}^{n} (Ax_i + Bv_i - 1)^2$$

as a measure of *closeness of fit*.

Box 7.1 **Choice of symbols for expressing results**

Many scientists give little thought to the symbols they use to present their ideas and results. This is a pity, because careful choice of symbols is always helpful to the reader and may also often be helpful as a way of avoiding mistakes.

Let us start with a non-mathematical example. At first sight, a non-biochemist might feel that (S)-1-phenylmethyl-1-aminoacetic acid, L-phenylalanine, phenylalanine, Phe, and F were equally good names or symbols for the same substance (provided that appropriate definitions are available for reference). In practice, however, they all have their uses and should not be interchanged arbitrarily. The first name would only be appropriate for a reference work for chemists where the aim is to specify the structure (including the stereochemistry) unambiguously. Biochemists would rarely use such a name, however, as they would take the view that 'everyone knows' what alanine is and that 'phenylalanine' is one in particular of the various compounds that could be regarded as phenyl derivatives of alanine. Moreover, except in special contexts, biochemists would omit the L- on the grounds that the stereochemistry could be assumed unless there is any reason to think otherwise. However, even phenylalanine is much too long for use in writing amino acid sequences, and it would be replaced by Phe in short sequences or F in long ones. Notice, however, that there is an important difference between these last two. It requires only a basic knowledge of protein chemistry to guess that Phe stands for phenylalanine, so this is a symbol that presents few difficulties for non-specialists. F is another matter, however, and unless one uses one-letter amino acid symbols all the time, some of the more obscure ones (not only F but others such as K for lysine) are not easy to recognize. It follows that one should show restraint in using symbols like F in work intended to be read by non-specialists.

(continued overleaf)

(continued)

All of the same considerations apply to the choice of mathematical symbols in biochemistry, but there are some additional ones as well. A symbol like $K_{i(Phe)}$ for an inhibition constant with respect to phenylalanine is sufficiently obvious to a biochemist reading about an enzyme inhibited by phenylalanine as to hardly need definition and, once it has been defined, it is certainly not necessary to provide frequent reminders of what it means. For deriving an equation, however, it is very cumbersome: in handwriting one quickly becomes tired of writing out its seven characters each time it is used, and excessively cumbersome symbols often result in mistakes. One might easily prefer something much shorter, such as K_F for private work, only reverting to the full symbol $K_{i(Phe)}$ when preparing work for someone else to read. Similarly, in the examples discussed in Section 6.8 one might well have preferred to write h (or even x) rather than $[H^+]$ for one's private work.

In the mathematical stages of an investigation, it is often useful to concentrate attention on the purely mathematical aspects by *deliberately hiding* the physical meanings of the symbols. In the example discussed in the text, writing K_m as A not only avoids the need for a subscript, but also helps to ensure that in deriving the result we shall not be influenced by any prejudices about where we might expect K_m to appear in an expression. Writing $1/V$ as B not only hides its physical meaning but also allows simpler algebra and calculus by putting the quantity of interest in the numerator and not in the denominator. At the end, of course, it is important to convert these symbols back into physically meaningful ones.

The aim is now to find values of A and B that make S a minimum. If there were only one parameter, this would be a simple problem in differentiation, of the sort we considered in Chapter 4. However, we have to find a minimum in S not only with respect to A, but also simultaneously with respect to B. As we can vary A and B independently—there is no relationship that specifies B if we know A, or vice versa—this is an exercise in partial differentiation, which we must do with respect to A and B in turn. While differentiating with respect to A, we treat B as a constant:

$$\frac{\partial S}{\partial A} = \sum [2x_i(Ax_i + Bv_i - 1)] = 2A\sum x_i^2 + 2B\sum x_i v_i - 2\sum x_i$$

and while differentiating with respect to B, we treat A as a constant:

$$\frac{\partial S}{\partial B} = \sum [2v_i(Ax_i + Bv_i - 1)] = 2A\sum x_i v_i + 2B\sum v_i^2 - 2\sum v_i$$

Two points should be noted about these expressions: first, as all of the summations are from $i=1$ to n, we can omit the limits from the summation signs without risk of ambiguity; second, in the first and second summations on the right-hand side

$2A$ and $2B$ (respectively) are factors of every term and so can be multiplied once each, after summing, rather than n times before.

For any value of B, we can minimize with respect to A by finding a value of A that makes the first partial derivative zero; conversely, for any value of A we can minimize with respect to B by finding a value of B that makes the second partial derivative zero. To minimize with respect to both parameters simultaneously, we must make *both* partial derivatives zero simultaneously. Let us define $\hat{A}$ and $\hat{B}$, respectively, as the values of A and B that satisfy this condition, i.e.

$$2\hat{A}\sum x_i^2 + 2\hat{B}\sum x_i v_i - 2\sum x_i = 0$$

$$2\hat{A}\sum x_i v_i + 2\hat{B}\sum v_i^2 - 2\sum v_i = 0$$

These are now a pair of simultaneous equations in $\hat{A}$ and $\hat{B}$ as unknowns, exactly analogous to those we considered in Chapter 6, apart from the fact that we now have rather complicated coefficients such as $\sum x_i^2$ instead of the simple constants we had before. We can, therefore, use the determinant method to write down the solutions without further algebra:

$$\hat{A} = \frac{\sum v_i^2 \sum x_i - \sum x_i v_i \sum v_i}{\sum x_i^2 \sum v_i^2 - (\sum x_i v_i)^2}, \qquad \hat{B} = \frac{\sum x_i^2 \sum v_i - \sum x_i v_i \sum x_i}{\sum x_i^2 \sum v_i^2 - (\sum x_i v_i)^2}$$

Finally, we can revert to the original symbols to give a result with a more obvious significance to biochemistry:

$$\frac{\hat{K}_m}{\hat{V}} = \frac{\sum v_i^2 \sum v_i / a_i - \sum v_i^2 / a_i \sum v_i}{\sum v_i^2 / a_i^2 \sum v_i^2 - (\sum v_i^2 / a_i)^2}, \qquad \frac{1}{\hat{V}} = \frac{\sum v_i^2 / a_i^2 \sum v_i - \sum v_i^2 / a_i \sum v_i / a_i}{\sum v_i^2 / a_i^2 \sum v_i^2 - (\sum v_i^2 / a_i)^2}$$

and it is a simple matter to rearrange these into expressions for $\hat{V}$ and $\hat{K}_m$:

$$\hat{V} = \frac{\sum v_i^2 / a_i^2 \sum v_i^2 - (\sum v_i^2 / a_i)^2}{\sum v_i^2 / a_i^2 \sum v_i - \sum v_i^2 / a_i \sum v_i / a_i}, \qquad \hat{K}_m = \frac{\sum v_i^2 \sum v_i / a_i - \sum v_i^2 / a_i \sum v_i}{\sum v_i^2 / a_i^2 \sum v_i - \sum v_i^2 / a_i \sum v_i / a_i}$$

These are now the 'best' values of V and K_m, in the sense that they make the function S that we defined as our criterion of closeness of fit as small as possible.

Before leaving this result, it is advisable to check it for algebraic errors with the dimensional criteria we examined in Chapter 2. As $\hat{V}$ is a rate, its expression must also be a rate: the first sum in the numerator is $\sum v_i^2 / a_i^2$, which must have the dimensions of v^2/a^2, because each individual term in the summation has these dimensions; the second sum, $\sum v_i^2$, must similarly have the dimensions of v^2; when these two are multiplied together the result must have the dimensions of v^4/a^2. The last sum in the numerator has the dimensions of v^2/a, and when this is squared the result has the dimensions of v^4/a^2, the same as those of the first product, as they must be if the subtraction is to be valid.

Exactly the same sort of analysis applied to the denominator shows that both products have the dimensions of v^3/a^2, so it is valid to subtract one from the other to give a denominator with the same dimensions of v^3/a^2. Finally, to calculate $\hat{V}$, we divide a numerator with dimensions of v^4/a^2 by a denominator with dimensions of v^3/a^2, so the result has the dimensions of v, exactly as expected for $\hat{V}$. The calculation is thus dimensionally consistent.

The same analysis applied to the expression for $\hat{K}_m$ and shows that both subtractions are valid and produce a numerator with dimensions of v^3/a and a denominator with dimensions of v^3/a^2, and that when one is divided by the other the result has the dimensions of a, i.e. the result is a concentration, as expected for $\hat{K}_m$. Thus, this calculation is also dimensionally consistent.

7.4 **Problems**

7.1 Determine partial derivatives for the function $z=(x^2+y^2)^{1/2}$ as follows:

(a) $\left(\dfrac{\partial z}{\partial x}\right)_y$,   (b) $\left(\dfrac{\partial x}{\partial y}\right)_z$,   (c) $\left(\dfrac{\partial y}{\partial z}\right)_x$.

7.2 For the function considered in problem (7.1), prove the following two relationships:

(a) $\left(\dfrac{\partial z}{\partial x}\right)_y\left(\dfrac{\partial x}{\partial y}\right)_z\left(\dfrac{\partial y}{\partial z}\right)_x=-1$,   (b) $\dfrac{\partial}{\partial y}\dfrac{\partial z}{\partial x}=\dfrac{\partial}{\partial x}\dfrac{\partial z}{\partial y}$

7.3 For the function $pV=RT$ (where p, V, and T are variables and R is a constant), show that

(a) $\dfrac{\partial V}{\partial T}\dfrac{\partial T}{\partial p}\dfrac{\partial p}{\partial V}=-1$,   (b) $\dfrac{\partial}{\partial p}\dfrac{\partial V}{\partial T}=\dfrac{\partial}{\partial T}\dfrac{\partial V}{\partial p}$

7.4 Consider a set of observations (x_i, y_i), for $i=1$ to n, that can be represented as values from the straight-line relationship $y_i=a+bx_i+e_i$, with additive deviations e_i. Defining the sum of squares as $S=\Sigma e_i^2$, find the values of $\hat{a}$ and $\hat{b}$ such that S is a minimum when $a=\hat{a}$ and $b=\hat{b}$.

7.5 Show that your solution to problem (7.4) is dimensionally consistent. (*Note*: Although no dimensions for x_i and y_i have been specified, dimensional analysis can still be used, as all that is necessary is to assume that the dimensions of x_i and y_i are different)

CHAPTER 8

Ideas of statistics

8.1 **Introduction**

In general, in this book I have tried to follow the principle that topics in mathematics should not be taught as recipes, i.e. if the theory is too advanced to be taught at a particular level, then the practice is also too advanced to be taught at that level. I have made one exception already, by suggesting that apart from a few simple and well known cases the most convenient way of integrating a function is to look it up in standard tables and, if this is done, one can use the result without any knowledge or understanding of how it was obtained. However, integration is a special case, because even if one cannot easily integrate an unknown function, it is not normally difficult to apply the process in reverse: given an expression that is supposedly the integral of some function, it is not difficult to differentiate it to check that it does indeed yield the original function.

This chapter will consider a different sort of exception: methods derived from statistics are very widely applied throughout experimental science, yet very few of the people who used them have any idea of their theoretical basis. Although the basic notions of probability that were examined in Sections 1.8 and 1.9 are not particularly difficult to understand, progressing from them to the theoretical basis of Student's t-test (one of the most widely used statistical methods used by experimental scientists), for example, is entirely another matter, one that is not covered even in elementary courses of statistics. Not surprisingly, therefore, it is only too easy to fall into the 'cookery book' style of teaching statistics, with the result that, when one looks for published examples of calculations that have been wrongly done, it is easier to find examples from statistics than from any other branch of mathematics.

The safest rule to follow is that statistics is just quantified common sense, and, if the results of a statistical calculation disagree with what is obvious, the calculation is more likely to be at fault. Unfortunately, this is not a rule that one is always allowed to follow. Journal editors, research directors, etc., sometimes insist that statements be supported by statistical tests, and experience shows that they often prefer absurd and meaningless statistical calculations over no statistical calculations at all. Therefore, whether one likes it or not, one cannot just say that statistical methods are too advanced to be taught in an elementary course and leave it at that.

8.2 Variation

In any kind of measurement, the result we obtain will not always be the same even if there is no reason to suspect that the property we are observing has changed between one observation and the next. Even if we are just counting the number of rats in a cage (a number that ought to remain absolutely constant as long as no rats are born and none die, and none are introduced into the cage or removed), the result may be 10 on the first count, 12 on the second, and 9 on the third, and so on, because when the objects to be counted are moving around it is difficult to be sure of not counting the same one twice or missing it altogether. This is an example of *measurement error*: the value we are trying to measure is exactly constant, but the actual measuring of it is liable to mistakes of one kind or another. Even if the actual measurement is done automatically with an instrument of some kind, measurement errors can (and nearly always do) arise because even the best instruments tend to drift a little in their calibration, or to be influenced by unpredictable variations in their environments, such as small changes in temperature or pressure, or the presence of varying amounts of dust in the atmosphere, etc.

In the physical sciences, measurement error may be all there is, at least for some kinds of measurements. For example, we normally assume that the temperature at which pure copper melts under specified conditions is a definite fixed value that does not vary with the weather or the time of day, etc., so that all of the variation we may encounter can be attributed to measurement errors. Many other quantities in the physical sciences may not be exactly constant, but any variation in their values may be trivial compared with measurement errors: this would be the case, for example, for the mass of a sample of rock, where any changes that result from slow oxidation or accumulation of dust particles, etc., may normally be ignored.

Matters are very different in the biological world. The masses of the rats in a cage will certainly not all be identical even if they came from the same litter of a pure inbred strain and have been maintained since birth under uniform conditions. Uncontrolled variations in access to food, temperature, and other unknown variables, will mean that some animals will be healthier and stronger than others. Even molecules of a protein cannot be assumed to be identical in sequence: the genetic apparatus is not perfectly accurate and, even with all the error-checking mechanisms that it contains, the final proteins inevitably have about 1 residue in 10^9 incorrect: this seems a small fraction, but it implies that in a protein of

500 residues around 1 in 2×10^6 molecules will be incorrect. Although some of these are likely to be so abnormal in properties that they will be eliminated from the cell, a great many will not. Moreover, even a small *proportion* of abnormal molecules adds up to a large *number*: a 1nM solution contains about 6×10^{14} molecules of solute per litre, so if 1 in 2×10^6 of these are abnormal, the total number of abnormal molecules in a 1 mL sample will be 3×10^5, by no means a negligible quantity.

Even if we ignore these considerations and believe that all of the molecules of a particular biological substance are identical while they are in their natural state in the living cell, they will not suffer precisely the same stresses during the process of extraction and purification. So, variations may arise that were not there at the beginning, and these will necessarily contribute to any variations in the kinetic measurements we may make on the purified substance.

Sometimes, biological variations of this kind may be large compared with the likely errors in measuring them, as with the masses of animals from a wild population; sometimes, they may be intermediate, as with the masses of animals from a uniform laboratory strain; sometimes, they may be small, as with the variations between the kinetic properties of molecules in a carefully purified sample of enzyme. However, the important point is that they will always exist and they contribute to the variation that is observed. In some circumstances, we shall be interested in knowing how much the biological variation is, and in others we shall prefer to combine it with the measurement error, regarding the total variation just as a departure from some ideal. Part of the task of statistics is to provide tools for distinguishing between these, but this is not necessarily easy, and so we shall begin with the simpler problem of how we can quantify the degree of variation in a set of measurements on the assumption that all of this variation reflects actual variation in the underlying truth, not measurement errors.

8.3 **Averages**

Suppose we have a sample of ten rats with masses 94, 115, 123, 88, 101, 87, 99, 107, 164, and 119 g, and we want to have some measure of their variability, assuming that all of the variation reflects actual differences in their masses and not measurement errors. The quickest and most obvious measurement is simply the range from the lightest to the heaviest, which we can determine by inspection to be 87–164 g. However, although this certainly tells us something we may want to know about the variability, it completely fails to convey the possibly important fact that all but one of the rats weighed less than 125 g. We are more likely to want to know how far on average a rat chosen at random is likely to differ from the average of all the rats. It is immediately clear then that, before we can begin to answer this question, we have to define a measure of what a statistician would call *central tendency* but what, in more ordinary language, we should call the *average*.

Although in everyday use there is just one kind of average, what we shall presently define as the *arithmetic mean*, one can easily think of several other possibilities. The simplest to determine, even for large numbers of values, is the *mid-range*, the value half way between the largest and smallest: for the sample of rats this would be $\frac{87+164}{2} = 125.5\,g$. Even if there were 1000 rats rather than 10, it would not take long to calculate this value, but it is not necessarily a very useful measure of the 'average rat', and we can notice that it is in fact larger than all but one of the individual values. This is because it is greatly affected by the presence of one or two abnormal values in an otherwise fairly homogeneous sample.

At the other extreme, we could choose a measure that is hardly affected at all by the presence of some extreme values, and we could notice that half of the values are 101 g or less, whereas half are 107 g or more. We could then define the *median* as the value 104 g that is half-way between these, which divides the sample into two equal halves. (Logically, we could take any value between 101 and 107 g as the median, but the convention for an even number of values is to define it as the value half-way between the two middle ones.) In contrast to the mid-range, the median is hardly affected at all by one or two abnormal values: if the 164 g rat weighed 850 g instead of 164 g, the median would still be 104 g, and if it weighed only 5 g, the median would be decreased only to 100 g. This resistance to extreme values can be quite useful for some purposes, but we may not want to ignore abnormal behaviour to this extent. In any case, the median has the additional disadvantage that it requires special methods of calculation and is not easy to calculate unless the sample size is very small.

For most purposes, therefore, we use an average that has properties intermediate between those of the mid-range and the median, known as the *mean*.

For the moment, we shall consider the commonest kind of mean, called the *arithmetic mean*, but we should mention that it is not the only possibility and that others exist. The arithmetic mean is the everyday 'average', obtained by adding up all the individual values and dividing by the number of values:

$$\bar{x} = \frac{1}{n} \sum_{i=1}^{n} x_i$$

$$\bar{x} = (94 + 115 + 123 + 88 + 101 + 87 + 99 + 107 + 164 + 119)/10 = 109.7\,g$$

The first of these equations is a definition of the arithmetic mean for a set of values x_i for $i = 1$ to n, and the second is an application of this definition to the set of rat masses. Notice that, unlike the mid-range, the mean requires some calculation for its evaluation but, unlike the median, it can be calculated in a very straightforward and mechanical way that lends itself to automatic implementation in the computer. Although it is not an absolute rule that the arithmetic mean is intermediate in value between the median (104 g) and the mid-range (125.5 g), it is not unusual that this should be so, because it is much more influenced by extreme values than the median, but much less so than the mid-range.

A fourth kind of average is the *mode*, which is the value that occurs most often. If all the values are different, then this definition has no meaning, but if we are willing to *group* the values and treat them as representative of a much larger sample, then we can count three values in the range 85–94 (87, 88, 94), two in the range 95–104 (99, 101), two in the range 105–114 (101, 107), two in the range 115–124 (119, 123), none in any of the ranges 125–134, 135–144, 145–154, and one in the range 155–164 (164). So the range in which the greatest number of values occur is the first, 85–94. To conclude from this that the same would be true if there were a much larger number of values would be very rash, as there is not nearly enough information to justify such a conclusion. However, it is the best we can do without examining a much more complex example than would be appropriate for a book of this level. We shall return to the mode when we come to the idea of a continuous *distribution* (Section 8.6) but, for the moment, we shall leave it as an inappropriate kind of average for studying small samples.

Therefore, returning to the mean, it is important to realize that not all of the values in a sample need be equally reliable, and so we may not want to give them all equal *weight*. Suppose, for example, that the numbers we started with were not masses of individual rats, but means of the masses of the rats in ten cages, and that the number of rats in the ten cages were not all the same. If, for example, the mean of 94 g for the first cage was derived from three rats, but the mean of 115 g came from 15 rats, we might want to take account of the fact that five $\left(\dfrac{15}{3} \right)$ times as much information was used to get the second mean as was used for the first, and we could do that by using a *weighted mean*:

$$\bar{x} = \frac{\Sigma w_i x_i}{\Sigma w_i}$$

Box 8.1 **Unweighted and weighted means**

The following are measurements of blood glucose levels (in mM) in a small clinical laboratory on three successive days, with different numbers of patients on each day:

Monday: 5.87; 6.21; 4.45; 7.56
Tuesday: 3.57
Wednesday: 6.66; 4.77; 5.17; 6.66; 4.99; 7.23

What is the mean value?

To answer this, we need to decide what is meant by the 'mean value'. If we just mean the arithmetic mean of all the measurements, it is

$(5.87 + 6.21 + 4.45 + 7.56 + 3.57 + 6.66 + 4.77 + 5.17 + 6.66 + 4.99 + 7.23)/11$

$= \dfrac{63.14}{11} = 5.74 \, \text{mM}$

(continued overleaf)

(continued)

If we mean the mean of the three daily means, then first we must calculate these:

Monday:
$$\frac{5.87 + 6.21 + 4.45 + 7.56}{4} = \frac{24.08}{4} = 6.02 \, \text{mM}$$

Tuesday:
$$\frac{3.57}{1} = 3.57 \, \text{mM}$$

Wednesday:
$$\frac{6.66 + 4.99 + 7.23}{6} = \frac{35.49}{6} = 5.91 \, \text{mM}$$

and then their mean:
$$\frac{6.02 + 3.57 + 5.91}{3} = \frac{15.51}{3} = 5.17 \, \text{mM}$$

These two answers are by no means the same, and the obvious explanation is that the day with the fewest measurements (Tuesday) also happened to give the lowest individual measurement. We may feel it more appropriate to *weight* each daily average according to the number of patients examined that day:

$$\frac{4 \times 6.02 + 1 \times 3.57 + 6 \times 5.91}{4 + 1 + 6} = \frac{24.08 + 3.57 + 35.46}{11} = \frac{63.11}{11} = 5.74 \, \text{mM}$$

This is just the number we obtained by taking the mean of all the individual measurements. This is not by chance, because by writing 4×6.02 we reverse the calculation of $\frac{24.08}{4}$ that we did to obtain the Monday mean in the first place, and although we do not recreate the individual values in $5.87 + 6.21 + 4.45 + 7.56$ by multiplying their mean by 4, we do recreate their sum. In a sense, therefore, calculating a weighted mean is a way of recovering some of the information that was lost when the original sets of values were converted to means for the individual days.

in which each w_i is the weight associated with each x_i. In the example, we might put each weight equal to the number of rats in the cage, so $w_1 = 3$, $w_2 = 15$, and so on. It is easy to show that the unweighted mean considered first is just a weighted mean with all weights equal, because if $w_i = w$ for all i then $\Sigma w_i x_i = w \Sigma x_i$, $\Sigma w_i = wn$ and w cancels to give the same expression as above.

Sometimes, there may be a reason to think that some values are *systematically* (which here really means predictably) more reliable than others. For example, an inhibition constant is (under specific conditions that do not need to be considered here) the concentration of an inhibitor that causes the rate of a reaction to decrease by half. It follows that if the inhibition is strong we can achieve this with very little inhibitor, but if it is weak then we need more; in other words, a large inhibition constant means weak inhibition. It follows that if a group of compounds

have inhibition constants of 1, 2, 2.5 and 150 mM, then the first three are reasonably good inhibitors, whereas the fourth is so weak that it is likely to be difficult to
measure the value with any accuracy. So, we may think that the first three values
are quite accurate, but for the fourth the real value could be anywhere from 100
to 500 mM. Yet the mean value will be 38.9, 26.4, or 126.4, depending on whether
we use 150, 100, or 500 mM for the highest value: clearly, it is tremendously
dependent on the least reliable value. It makes more sense—as long as we agree
that it is easier to measure strong inhibition than weak inhibition—to convert the
inhibition constants, which are measurements of weakness, into measurements of
strength, average those and then convert back. In other words, we can calculate
the reciprocal of the mean reciprocal. This is called the *harmonic mean*: it is defined
in general as

$$\bar{x} = \frac{n}{\Sigma(1/x_i)}$$

and application to the specific example gives

$$\frac{4}{\dfrac{1}{1}+\dfrac{1}{2}+\dfrac{1}{2.5}+\dfrac{1}{150}} = \frac{4}{1+0.5+0.4+0.0067} = \frac{4}{1.9067} = 2.10 \, \text{mM}$$

It is a simple matter to confirm that this result is hardly altered by putting 100
or 500 mM instead of 150 mM.

In keeping with the idea that we use the harmonic mean when it is easier to
measure small values than large ones, it is normally only applied to types of measurement where zero and negative values are unattainable. It makes little sense
to calculate the harmonic mean of a sample that includes zero or negative values.
In the case of the inhibitor example, an inhibition constant of zero would mean
the rate was zero even in the absence of the inhibitor, possible, of course, if the
'enzyme' did not in fact catalyse the reaction, but nothing to do with the identity
of the particular inhibitor that was absent.

One last kind of mean to consider is the *geometric mean*, which can be regarded
either as the nth root of the product of all the values:

$$\bar{x} = \left(\prod x_i\right)^{1/n}$$

or as the antilogarithm of the arithmetic mean of their logarithms:

$$\bar{x} = \exp\left(\frac{\Sigma \ln x_i}{n}\right)$$

It should be easy to show that these are exactly equivalent if all the x_i values are
positive. If some of them are zero or negative, then the second definition cannot
be applied and the first may or may not be usable. However, this distinction has
no practical importance as the geometric mean should only be used if all the values are positive. Indeed, it is particularly appropriate in cases where all of the

values *must* be positive, as is the case with many of the quantities that need to be measured in science: for example, the mass of a rat, the Michaelis constant of an enzyme, the concentration of a metabolite, etc., for which no meaning can be applied to a negative value. As noted in Box 8.2, geometric means are actually used much more often in biochemistry and other sciences than may be obvious.

Box 8.2 **The geometric mean: more often used than we may think**

It is quite rare for a biochemist to calculate a geometric mean explicitly (more rare than it ought to be, given that many biochemical measurements satisfy the criteria for a geometric mean to be appropriate, as they are constrained to be positive numbers). Nonetheless, geometric means are often used *implicitly*, whenever the values are commonly considered in logarithmic form.

Consider four samples of an assay mixture for which a pH meter showed pH values of 7.48, 7.51, 7.56, and 7.47. What is the average hydrogen-ion concentration for the four samples? Calculation of the arithmetic mean would require converting the four pH values to concentrations, adding them together and dividing by 4:

$$\text{Mean } [H^+] = \frac{10^{-7.48} + 10^{-7.51} + 10^{-7.56} + 10^{-7.47}}{4}$$

However, in practice this is unlikely to be what one would do. It would be more usual to calculate the arithmetic mean of the pH values:

$$\text{Mean pH} = \frac{7.48 + 7.51 + 7.56 + 7.47}{4}$$

and convert this to the hydrogen-ion concentration at the end.

As an exercise, you should satisfy yourself, by following through the calculations both algebraically and in terms of the numbers given, that this latter calculation will actually give the geometric mean of the concentrations, not the arithmetic mean.

8.4 **Measures of dispersion**

We can now return to the question posed in Section 8.2: given a set of values that are not all same, how can we quantify the amount of variation? If we take the arithmetic mean as a convenient measure of average (without assuming that it is necessarily any better than any of the other kinds of averages discussed in Section 8.3), it seems natural to take the average difference between the individual values and the mean as the measure of their dispersion. Thus, using the same set of numbers (94, 115, 123, 88, 101, 87, 99, 107, 164, and 119 g, with mean 109.7 g) used already as example, we might calculate

$$[(94-109.7)+(115-109.7)+(123-109.7)+(88-109.7)+(101-109.7)$$
$$+(87-109.7)+(99-109.7)+(107-109.7)+(164-109.7)+(119-109.7)]/10$$
$$=(-15.7+5.3+13.3-21.7-8.7-22.7-10.7-2.7+54.3+9.3)/10$$
$$=0/10=0$$

Clearly, something has gone seriously wrong, as apparently we find no dispersion even though it is obvious from inspection that the values are not all the same. The problem is that the negative values exactly balance the positive ones in this calculation and, if we remember how the mean of 109.7 was calculated, it should be clear that this will always happen and that it is not a special property of the particular set of numbers used.

Conceptually, the simplest way to remedy the problem is just to ignore the signs, treating all the differences as positive, and doing this produces the *mean deviation*:

$$(15.7+5.3+13.3+21.7+8.7+22.7+10.7+2.7+54.3+9.3)/10=16.44\,\text{g}$$

This is a reasonable answer to the question of how much the individual numbers differ on average from the mean, but in practice the mean deviation is seldom used. This is not so much because it is a poor measure but because just ignoring signs is difficult to handle algebraically, so that deriving general properties of the mean deviation is a rather advanced topic.

It proves to be a great deal simpler to eliminate the negative signs by squaring all the values to get the *mean squared deviation*, which is given a special name, the *sample variance*, and symbol, s^2:

$$s^2=[(-15.7)^2+5.3^2+13.3^2+(-21.7)^2+(-8.7)^2+(-22.7)^2$$
$$+(-10.7)^2+(-2.7)^2+54.3^2+9.3^2]/10=467.01\,\text{g}^2$$

Box 8.3 **A more convenient formula for the sample variance**

The sample variance is defined as $s^2=\frac{1}{n}\Sigma(x-\bar{x})^2$. Although this is conceptually the simplest way to think about it—as the mean of the squared deviations from the mean—it is not the most convenient definition to use for calculating it. Expanding the square, and recalling the definition of $\bar{x}$, we can write $ns^2=\Sigma(x^2-2x\bar{x}+\bar{x}^2)=\Sigma x^2-2\bar{x}\Sigma x+\Sigma\bar{x}^2$. In this sum the meaning of the first term is clear enough—it is just the sum of all the individual x^2 values. The second is $-2\bar{x}$ times Σx, and as Σx is just $n\bar{x}$ (from the definition of $\bar{x}$) this is just $-2n\bar{x}\bar{x}$, or $-2n\bar{x}^2$. What of the third term, $\Sigma\bar{x}^2$, however? As $\bar{x}^2$ is a constant, equal to $(\Sigma n)^2/n^2$ this third summation requires us to add n instances of $\bar{x}^2$ together, and as with any other constant the result is just n times the constant, i.e. $n\bar{x}^2$. So the sum of three terms can be written as $ns^2=\Sigma x^2-2n\bar{x}^2+n\bar{x}^2$, or more simply, as $ns^2=\Sigma x^2-n\bar{x}^2$. Thus the sample variance now takes the

(continued overleaf)

(continued)

following form:

$$s^2 = \frac{1}{n}\sum x^2 - \bar{x}^2$$

This is much more convenient for calculation, as it requires far fewer subtractions, and in practice it is usual to calculate the mean and sample variance simultaneously by setting up a table; x and x^2 are written in two columns and their sums calculated together.

To illustrate this with a simple example, let us calculate the mean and standard deviation of the numbers $(1.2, 5.3, 3.1)$. In the long-winded calculation, we first calculate $\Sigma x = 1.2 + 5.3 + 3.1 = 9.6$, and divide by 3 to get $\bar{x} = \frac{9.6}{3} = 3.2$. We then calculate $(1.2 - 3.2, 5.3 - 3.2, 3.1 - 3.2) = (-2, 2.1, -0.1)$, and square them to get $(4.00, 4.41, 0.01)$. Adding these gives $3s^2 = 8.42$, hence $s^2 = \frac{8.42}{3} = 2.807$, and $s = 1.675$. In all, we have done four additions, three subtractions, two divisions, as well as taking three squares and one square root.

If now we repeat the calculation as a table, we have

x	x^2
1.2	1.44
5.3	28.09
3.1	9.64
9.6	39.17

Hence $\bar{x} = \frac{9.6}{3} = 3.2$, $s^2 = \frac{39.17}{3} - 3.2^2 = 2.817$, $s = 1.675$. We still needed four additions (all done in one series of operations, rather than as two sets of two), two divisions and one square root, but four squares (instead of three) and only one subtraction (instead of three). Even in this simple example there is a perceptible saving of effort. More generally, with a sample of n one saves $n - 1$ subtractions with a cost of one extra square, which can add up to a substantial saving of labour.

The shorter calculation has the additional benefit that doing most of the calculation in a table makes it both easier to avoid mistakes in the first place, and easier to check the workings afterwards. Set against this, however, we must put the fact that using the short calculation requires a new formula to be memorized, whereas the longer one requires only a knowledge of how a standard deviation is defined. Thus, one can always rely on the longer method in the event that one is unsure of the formula for the shorter one.

For ordinary use (i.e. other than in theoretical statistics), it is more convenient to consider the square root of this quantity, which is called the *sample standard deviation*, because it has the same dimensions (mass in this case) as the values to which it refers:

$$s = 21.6\,\mathrm{g}$$

This now provides a convenient measure of how much the individual values in the sample vary among one another, and is the most widely used measure for that purpose.

8.5 **The idea of a population**

The particular set of values for which we measure the variance or standard deviation may be the entire set that is of interest to us. This would be the case, for example, if we were analysing examination results and the analysis included the results for the entire set of students who had taken the examination. In this case, we could very well be interested in the dispersion of results for those particular students and we would not regard them in any way as 'representing' a much larger group of students who *might* have taken the same examination. We then make no distinction between sample and population, because the sample *is* the population, and there is no reason to look any further than the sample standard deviation or the variance as a measure of dispersion.

Perhaps surprisingly, this is not the usual case in science. If we measure the masses of a group of rats, it is unlikely that we are truly interested in those particular rats as individuals. It is much more likely that we want to know how much variation in mass to expect in 'typical' rats at a particular age from an inbred line submitted to a particular dietary regime. In this case, we regard the rats weighed as a *sample* from an infinite *population* consisting of all the rats with the right characteristics that might have been studied. This may seem rather abstract, but it is a fundamental idea that underlies much of statistical theory and practice. It applies not only to individual objects like rats but also to more intangible things like experiments: if we make ten measurements of the pH of a buffer, they constitute a sample of the infinite measurements we might have made in the same way.

As random errors tend to cancel out as a sample gets larger, we are often interested in trying to judge what the statistical properties of a sample would be if we made the sample size infinite. It might seem that we could just use the sample mean and standard deviation as the most appropriate measures we could find of the corresponding population values. In the case of the mean, the sample mean is indeed an appropriate estimate of the population mean, and is the one that is used nearly always, but there is a subtle complication that needs to be taken account of for the standard deviation. Let us consider the most extreme case possible in which the sample consists of just one rat, weighing 94 g. Are we entitled to assume that the average rat weighs 94 g? Maybe not (certainly not if we consciously biassed the choice of rat by taking the one that looked smallest to the eye, and probably not if we unconsciously biassed it by taking the one that caught the eye first), but in the absence of any other information it is the best we can do.

The sample standard deviation for a sample of one rat must be zero, as it is just the difference between the one value, 94 g, and the mean, which is also 94 g. But do we have any justification for assuming that there is no variation in mass in the whole set of rats? Clearly not, as all our experience tells us that there must be some variation. Thus, whatever the true value of the standard deviation, it is certainly greater than zero. If there were two rats, weighing 94 and 115 g, the sample mean would be different, $\frac{94+115}{2} = 104.5$. It would be fair to say that this is likely to be a 'better' estimate of the population mean than 94 g, in the sense that it is based on more information, and it allows us to calculate a non-zero standard deviation:

$$s = \sqrt{\{[(94-104.5)^2 + (115-104.5)^2]/2\}} = 10.5 \, g$$

However, do we have any more reason to believe that it is a good measure of the population standard deviation? It is still based on very little information, and notice that we have biassed the calculation by choosing not just any value as the estimate of the mean but the particular value 104.5 that makes the two differences equal (apart from sign), with the smallest sum possible. So, we have, in effect, defined the mean in such a way that the standard deviation is as small as we could make it. Thus, in general we must still expect it to be an underestimate. How much of an underestimate is not obvious, and we need not go into details here. It turns out that the best estimate of the *population variance*, which we now represent as σ^2 to distinguish it from the sample variance s^2, is obtained by multiplying the sample variance by $n/(n-1)$:

$$\sigma^2 \approx \frac{ns^2}{(n-1)}$$

and so the best estimate of the population standard deviation is just the square root of this:

$$\sigma \approx s \sqrt{\frac{n}{(n-1)}}$$

The population standard deviation tells us something about the variability of the numbers that go to make up our sample, but this may not necessarily be what we want to know. If we have measured the arithmetic mean in the hope of learning something about the arithmetic mean of the population, then we shall certainly want to have some idea of how good a measure it is. In other words, we want to have some idea of the variability of the (calculated) sample mean as an estimate of the (unknown) population mean. It is intuitively obvious that the sample mean ought to get better and better as an estimate of the population mean as the sample gets bigger, but that is not how the estimated population standard deviation behaves: as the sample gets bigger, it gets closer and closer to the true population standard deviation, which is not zero but a definite positive value. This suggests that the estimated population standard deviation is too big if we are interested in a measure of the precision of the mean, and that we ought to adjust it downwards as the sample gets bigger, to take account of the increased amount of information that it contains. Again, without going into details of the derivation, we may just consider the result of these considerations, which is that $\sigma^2(\bar{x})$, the estimated variance of the mean, is smaller by a factor n than $\sigma^2(x)$, the estimated variance of the population, i.e.

$$\sigma^2(\bar{x}) \approx \frac{\sigma^2(x)}{n}$$

As the latter quantity is the variance just written as σ^2 above, it follows that the formula for the variance of the mean is

$$\sigma^2(\bar{x}) \approx \frac{s^2}{(n-1)}$$

The square root of this is, naturally enough, an estimate of the standard deviation of the mean. However, the term commonly used is not that but *standard error of the mean*, often abbreviated to s.e.m.:

$$\text{s.e.m.} = \frac{s}{\sqrt{(n-1)}}$$

....

Example 8.1 Averages and measures of dispersion

The following set of values are measurements of lengths (in μm) of five different hepatocytes observed under a microscope: $(31.2, 35.7, 40.2, 33.1, 28.6)$. Calculate the following values (including units in all cases): (a) the sample size, n; (b) the arithmetic mean, $\bar{x}$; (c) the geometric mean; (d) the harmonic mean; (e) the median; (f) the mid-range; (g) the sample variance, s^2; (h) the sample standard deviation, s; and estimate the following values (*also with units*): (i) the population variance, σ^2; (j) the population standard deviation, σ; (k) the standard error of the mean; (l) the population mode.

(a) The sample size is the number of observations, i.e. $n = 5$ (no units).

(b) The arithmetic mean $\bar{x} = \dfrac{31.2 + 35.7 + 40.2 + 33.1 + 28.6}{5} = 33.76\,\mu$m.

(c) The   geometric   mean $= (31.2 \times 35.7 \times 40.2 \times 33.1 \times 28.6)^{\frac{1}{5}} = 33.53\,\mu$m. This can also be calculated by calculating the arithmetic mean of the logarithms, i.e. $\dfrac{\ln 31.2 + \ln 35.7 + \ln 40.2 + \ln 33.1 + \ln 28.6}{5} = \ldots$ and taking the exponential of the result: $\exp() = 33.53\,\mu$m.

(d) The harmonic mean is the reciprocal of the mean of the reciprocals, i.e.

$$\frac{5}{\frac{1}{31.2}+\frac{1}{35.7}+\frac{1}{40.2}+\frac{1}{33.1}+\frac{1}{28.6}}$$

$$=\frac{5}{0.03205+0.02801+0.02488+0.03021+0.03497}=33.31\,\mu m.$$

(e) Arranging the values in order gives (28.6, 31.2, 33.1, 35.7, 40.2); the median is then the middle value, i.e. $33.1\,\mu m$.

(f) The smallest value is $28.6\,\mu m$, the largest is $40.2\,\mu m$; the mid-range is therefore $\frac{28.6+40.2}{2}=34.4\,\mu m$.

(g) The sum of squared deviations from the mean is $(31.2-33.76)^2+(35.7-33.76)^2+(40.2-33.76)^2+(33.1-33.76)^2+(28.6-33.76)^2=78.85\,\mu m^2$, and dividing this by the sample size gives the sample variance, $s^2=15.77\,\mu m^2$.

(h) The sample standard deviation is the square root of the sample variance, $s=15.77^{0.5}=3.97\,\mu m$.

(i) The number of degrees of freedom after calculating the mean is one less than the sample size, so the population variance is estimated as $\sigma^2=\frac{78.852}{4}=19.71\,\mu m^2$.

(j) The population standard deviation is the square root of the population variance, so it is estimated as $\sigma=19.71^{0.5}=4.44\,\mu m$.

(k) The standard error of the mean is obtained by dividing the sample standard deviation by the square root of the sample size minus one, i.e. s.e.m. $=\frac{3.97}{(5-1)^{0.5}}=\frac{3.97}{2}=1.99\,\mu m$.

(l) It is impossible to estimate a meaningful value of the population mode from so small a sample. However, if the sample size were considerably larger than 5 one might estimate the mode from the arithmetic mean and the median as $3\times33.1-2\times33.76=31.8$.

Here we replace the approximate equality sign $\approx$ with an ordinary equality sign $=$, because this is a definition of the standard error of the mean, which is an estimate of the true population standard deviation of the mean. The term *standard error* is applied not only to means but to any of the values that we may calculate from experiments; for example, to kinetic constants. However, we shall not consider here how it is calculated in such more complex cases.

Notice that the standard error of the mean behaves as we would logically expect it to behave, in the sense that it gets smaller as the sample size gets bigger, approaching zero as n approaches infinity, reflecting our intuitive expectation that if we based our mean on an infinite sample we ought to know it with infinite accuracy, i.e. with zero error. This result has an important practical implication which applies, at least approximately, to all experimental science: if we want to decrease the experimental uncertainty in a measurement by a factor α, then we can only do so with the same experimental approach by increasing the number of measurements by a factor α^2. As α may often be 10 or more, this implies an increase of 100-fold or more in the experimental effort, which will rarely be feasible and will sometimes be impossible. As statistical theory offers no way of escaping this impasse, it provides a powerful incentive to searching for better experimental approaches. In the history of science, major advances have nearly always come from the development of more powerful or more accurate methods, and not from increased effort in the application of existing methods.

8.6 **Continuous distributions**

The simplest example of a *distribution* is provided by a histogram, which is just a way of showing how many of a set of measurements yielded values in each of several ranges. Suppose, for example, that we have measured the lengths of 48 cells under the microscope. We might represent the set of measurements with a histogram such as the one in panel (a) of Fig. 8.1 if we used ranges of 1 μm in width, or the one in panel (b) with ranges of 0.05 μm. Notice how the 'raggedness' of the data, i.e. the tendency of the successive strips to vary haphazardly in height, increases if we use narrower strips. This tendency increases further if the strips are made narrower still, as in panel (c). This is a general property of histograms: even though we may hope to observe the fine structure of the data by using a larger number of narrower strips, this just increases the noise. In practice, any histogram is a compromise between showing as much of the variation of interest as possible and as little noise as possible: too broad and the noise disappears but so does the information about variation; too narrow and the information about the variation is lost in the noise.

There is, however, a way to decrease the amount of noise enough to allow narrower panels and that is to measure more values: panels (d–f) show the sort of results one might have with 96 measurements instead of 48. Notice that the quality of the histograms in panels (a) and (e) is roughly the same, illustrating the general point that, to be able to halve the width of the strips without degrading the quality of the information, we require twice as many observations.

This tendency to smooth out the outline of the histogram as more measurements are included continues indefinitely and, if the number is very large, the irregularities almost disappear. It is not possible, of course, to include infinitely many measurements, but we can imagine that, if it were possible, the outline of the histogram would resemble the smooth curve that is superimposed on panel (e). This now represents the *continuous distribution* corresponding to the various *discrete distributions* illustrated by the histograms.

Because a continuous distribution refers to an infinitely large population, it is no longer necessary for the strips that compose the histogram to have a finite width; instead, we can imagine an infinite number of infinitely thin strips, and

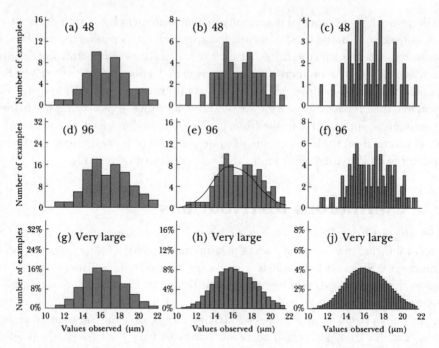

Fig. 8.1 Histograms representing the frequencies of different measured values. The curve representing the theoretical distribution is superimposed on panel (c). Reading the panels from left to right shows that the noise is increased by using narrower measurement ranges, and reading them from top to bottom shows that it is decreased by increasing the number of independent measurements.

counting frequencies now becomes a matter of integrating the area under any selected region of curve.

Although many different shapes of curve are possible, it is quite common to find a distribution that resembles a symmetrical (deviations of any given size from the mean are just as likely to be positive as to be negative) bell-shaped curve that is known as a *normal distribution* or a *Gaussian distribution*. This curve is illustrated in Fig. 8.2. It has a maximum at an observed μ, the *mean* of the distribution, and becomes very small for values greater than $\mu + 3\sigma$ or less than $\mu - 3\sigma$, where σ is the *standard deviation* of the distribution. These are often called the *population mean* and *population standard deviation*, respectively, and the similarity of these names to the sample mean and sample standard deviation that we have considered already is not by chance: if we take a sample from a normal distribution, we expect to get a sample mean close to the population mean, and a sample mean close to the population standard deviation. How close depends, of course, on the size of the sample: for three or four measurements in the sample, they may disagree quite badly, but as the sample size approaches infinity the agreement approaches exactness.

The ordinate value of the normal curve is not important for understanding it; what is important is the area underneath it, or its integral. As marked on the curve,

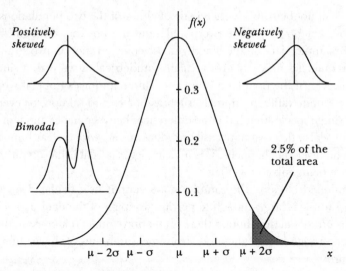

Fig. 8.2 The normal distribution curve. The x scale is calibrated in units of σ (the standard deviation) around the population mean μ. The y scale is calibrated in units such that the total area under the curve is 1 if $\sigma = 1$. Approximately 2.5% of this area lies beyond an x value of $\mu + 2\sigma$. The insets show various kinds of deviation from the normal distribution. Positively and negatively skewed curves have modes (the values of x at the maximum) smaller and larger than the mean, respectively. A bimodal curve has two modes, i.e. two maxima.

about 2.5% of this area lies beyond an abscissa value of $\mu + 2\sigma$, or about 5% if we add to this the equivalent area below $\mu - 2\sigma$. This is the basis for the common statement that 95% of observations lie within two standard deviations of the mean. This is true, at least approximately, but it should be interpreted with caution. First, it is a statement about the normal distribution, and although we may well have reason to think that a set of values are distributed approximately as a normal distribution, we can never be sure that they are exactly distributed in any particular way. Second, it is a statement about the relationship between observed values and the population mean and standard deviation, whereas all we ever have in practice are the sample mean and standard deviation. We do not need to discuss the implications of this here: it is sufficient to understand that statistical statements derived from the normal distribution can only be approximate statements about real data and that, in general, large deviations from expected behaviour occur in real life more often than the normal distribution may suggest.

Figure 8.2 also illustrates some other continuous distributions. If large positive deviations are more probable than large negative deviations, the distribution is said to be *positively skewed*; in the opposite case, it is *negatively skewed*. If the curve has two maxima, it is *bimodal* (in other words, it has two modes, if we recall from Section 8.3 that the mode is the value that occurs most often; for a distribution curve it is the abscissa value at the maximum). A bimodal distribution is the extreme result of mixing two populations. If we mix data from two populations, the results will not be truly characteristic of either of the two populations, but this may not be very obvious if the two populations are not very different from each other (for example, data on the blood-glucose concentration in two strains of rats) and the frequencies may still give a clean symmetrical curve with a single maximum. However, if the populations are very different, for example if we combine data for two quite different animals, such as horses and salmon, or even for two sexes in a single species that has large differences between males and females (such as elephant seals), then we may easily produce results with a clear bimodal distribution. Whenever this happens, it is a clear sign that data that ought to be kept separate are being mixed together.

As mentioned already, the sample mode cannot be calculated meaningfully unless the sample is large enough to permit grouping of the data. However, for a continuous unimodal distribution there is an approximate relationship that can be used to get a rough idea of the mode from measurements of the arithmetic mean and median:

$$\text{mean} - \text{mode} \approx 3 \times (\text{mean} - \text{median})$$

and hence

$$\text{mode} \approx 3 \times \text{median} - 2 \times \text{mean}$$

Notice that these relationships imply that the median should be between the mode and the mean, but much closer to the mean, in agreement with the order of the words in the dictionary: 'median' is between 'mean' and 'mode', but much closer to 'mean' than to 'mode'.

8.7 **Thinking statistically**

Perhaps the most important lesson to be learned from a study of simple statistics and probability is the usefulness of acquiring the habit of thinking statistically. This means thinking about any observation in terms of probability, not to try to do a quantitative calculation but just to get a rough idea of whether the pattern observed is surprising or not.

We can illustrate this with a simple example. The sequence of the human protein cytochrome c is as follows (written in the one-letter code):

GDVEKGKKIF IMKCSQHTVE KGGKHKTGPN LHGLFGRKTG QAPGYSYTAA
NKNKGIIWGE DTLMEYLENP KKYIPGTMKM IFVGIKKKEE RADLIAYLKK
ATNE

Suppose you are examining this sequence and you notice that it contains nine examples of pairs of identical residues in sequence (I have made it easy to notice this by underlining the pairs but, even without any underlining, it would not be difficult to detect just by running your eye across the sequence). For the moment, let us ignore the fact that five out of the nine cases are pairs of lysine residues (KK) and ask the question in the simplest form: how surprised ought we to be that a protein of a little more than 100 residues (104 to be exact) contains nine pairs of identical residues? Is this an interesting observation for which we should search for an explanation in the function of cytochrome c, or is it more or less what we should expect even if the sequence were completely random?

This sort of question has great importance in the study of protein and nucleic acid sequences, because many discoveries are the result of noticing patterns that appear to be surprising. Many computer programs now exist to help people decide whether what they noticed is interesting or not, but here we are concerned with the sort of rough idea we can get just from applying commonsense statistical thinking to the problem.

As 20 different kinds of amino acids occur in ordinary proteins, we can start with the rough idea that each of them has a 1 in 20 chance, or 5%, of occurring in any particular position. (We shall see in a moment how we can replace this 5% estimate with a more accurate one, but 5% is good enough to start the discussion.) It follows that whatever the amino acid may be at any particular position, there is about a 5% chance that the *same* amino acid will occur at the *next* position. It follows that there is a 5% chance that residues 1 and 2 will be the same, a 5% chance that residues 2 and 3 will be the same, and so on; but it is better to say that there is a 0.95 chance that residues 1 and 2 will be *different*, a 0.95 chance that

....

Example 8.2 Thinking statistically

Twenty-five people in a room discover on comparing birthdays that two of them were born on the same day of the year. Is this a surprising observation?

Ignoring the 29th February and assuming that all other days are equally likely, the probability that two birthdays at random are the same is $\frac{1}{365}$, so the probability that they are different is $\frac{364}{365}$: there are 365 days on which the second birthday could fall, of which 364 are different from the first one. For the third birthday, there are 363 ways of choosing a different day from both of the first two, so the probability that all three are different is $\frac{364 \times 363}{365 \times 365}$. For four, we have $\frac{364 \times 363 \times 362}{365 \times 365 \times 365}$, and as one. By the time we reach 25 the probability that they are all different is $\frac{364 \times 363 \times 362 \times \cdots \times 341}{365^{24}}$ or 0.431. Thus the observation is not surprising at all, but corresponds with what we ought to expect more often than not.

....

residues 2 and 3 will be *different*, and so on. Multiplying these together we can estimate a chance $0.95 \times 0.95 = 0.9025$, or about 0.9, that residues 1 and 2 are different *and* residues 2 and 3 are different. Thus, there is a probability of 90% that we shall see no pair of identical residues among the first three residues. However, human cytochrome c contains 104 residues, so there are 103 places where we could make the comparison, and we should calculate $0.95^{103} = 0.0051$, or 0.5%, as the probability that nowhere in the entire sequence of cytochrome c is there a pair of identical residues. Notice that considering the whole sequence instead of just one pair has converted a high probability of seeing no identity into a near certainty of seeing an identical pair somewhere.

This gives us $1 - 0.0051 = 0.9949$ as the chance of seeing *at least one* pair of identical residues. To calculate the chance that there is *exactly one* such pair, we must multiply 0.0051 by $\frac{0.05}{0.95}$ to allow for the probability that *any one* pair is identical, and again by 103 to allow for the fact that there are 103 places where this identity could arise, giving 0.0276. To calculate the chance that there are exactly two pairs, we multiply again by $\frac{0.05}{0.95}$ (for the same reason), by 102 to allow for the fact that there are 102 other places available for the second identity (not counting the one already taken), and we must divide by 2 to allow for the fact that it does not matter in which order we consider the two pairs, so we have 0.0742. Tabulating these numbers and continuing the calculation to higher numbers of pairs, we have the set of probabilities illustrated in Table 8.1.

In each row, the second number is the probability of obtaining that number of pairs, the third is the cumulative probability, i.e. the probability of obtaining that number or fewer, and the expression in the right-hand column is what we need to multiply by in order to obtain the probability in column 2 of the next row.

Looking now for the row that corresponds to the number actually observed in the cytochrome c sequence, we see that there is a chance of 0.0397 of finding

Table 8.1 Probabilities of observing specified numbers of pairs of identical residues in a sequence of 104 amino acids.

Number of pairs	Probability	Cumulative probability	Factor for the next line
0	0.0051	0.0051	$(0.05 \times 103)/(0.95 \times 1)$
1	0.0276	0.0327	$(0.05 \times 102)/(0.95 \times 2)$
2	0.0742	0.1069	$(0.05 \times 101)/(0.95 \times 3)$
3	0.1315	0.2384	$(0.05 \times 100)/(0.95 \times 4)$
4	0.0976	0.3360	$(0.05 \times 99)/(0.95 \times 5)$
5	0.1803	0.5163	$(0.05 \times 98)/(0.95 \times 6)$
6	0.1550	0.6713	$(0.05 \times 97)/(0.95 \times 7)$
7	0.1130	0.7843	$(0.05 \times 96)/(0.95 \times 8)$
8	0.0714	0.8557	$(0.05 \times 95)/(0.95 \times 9)$
9	0.0397	0.8954	$(0.05 \times 94)/(0.95 \times 10)$
10	0.0196	0.9150	$(0.05 \times 93)/(0.95 \times 11)$

The value of 0.0051 in the first line is 0.95^{103}, the probability that nowhere in the entire sequence is there a pair of identical residues, i.e. 0.95, assumed to be the probability that any given pair are different, raised to the power of the number of pairs to be considered. This is multiplied by the factor shown in the right-hand column in order to obtain the value of 0.0276 in the next line, and so on. See the text for a fuller explanation.

exactly nine identical pairs, and, one row above, 0.8557 of observing eight or fewer. This means there is chance of $1 - 0.8557$, i.e. 0.1443, or about 14%, of observing at least nine. This (14%) is not very small, so we should not be especially surprised to see nine identical pairs in the human cytochrome c sequence. Following the trend of the numbers, we can see that we would need to observe appreciably more than ten pairs, say 15 or 20, before we were sufficiently surprised to feel there was an interesting property deserving further study.

....

Example 8.3 An exercise in simple algebra

Prove that $(x-d)^2 + (x+d)^2$ is always at least as large as $2x^2$. (This is, of course, just an exercise in elementary algebra; it is included here to illustrate the generality of the argument that the more unevenly distributed are the amino acids the greater the chance that two randomly selected sites will contain the same amino acid.)

Multiplying out $(x-d)^2 + (x+d)^2$ we have $x^2 - 2dx + d^2 + x^2 + 2dx + d^2 = 2x^2 + 2d^2$, which exceeds $2x^2$ by $2d^2$, a positive (or more exactly a non-negative) quantity regardless of the sign of d.

As a more general and only slightly more difficult exercise, prove that $(x-d)^2 + (y+d)^2$ must be larger than $x^2 + y^2$ if y is larger than x and d is positive (which means the change is in such a direction as to increase whatever difference between y and x exists initially).

....

Can we use our knowledge of protein chemistry to refine this calculation? We assumed at the outset that all 20 amino acids were equally common, but this is not true: with only a superficial knowledge of protein chemistry, we should know that some amino acids, like alanine (A) and glutamate (E), are common, whereas others, like tryptophan (W) and methionine (M), are not. How does this affect the estimate of 0.05 as the probability that a randomly selected pair of positions will contain the same amino acid? This may not be immediately obvious, but we can see what the trend must be by considering a pair of amino acids such that one (say tryptophan, W) occurs in somewhat fewer than 5% of positions, say 4%, and the other (say alanine, A) correspondingly more, 6%, all of the others occurring at 5%, as before. It follows that the proportion of WW pairs is decreased from 0.05^2 to 0.04^2, or from 0.0025 to 0.0016, and the proportion of AA pairs is increased from 0.05^2 to 0.06^2, or from 0.0025 to 0.0036. This gives a sum of WW and AA pairs of $0.0016 + 0.0036 = 0.0052$, which is greater than the sum $0.0025 + 0.0025 = 0.0050$ that would apply if both amino acids had frequencies of 0.05. It turns out to be *true in general* that for any change in frequencies away from equality, the effect of increasing one frequency more than outweighs the effect of decreasing another, and so the more unevenly distributed the amino acids are the greater the chance of finding two randomly chosen positions (not necessarily adjacent) occupied by the same amino acid. For the distributions that typically occur in proteins, the realistic value is about 0.07: this differs from 0.05 by enough to

make pairs of identical residues noticeably more common than the above calculation would suggest. If you take the trouble to repeat the above calculation with
0.07 instead of 0.05, you should find that the probability of observing nine or more
identical pairs is about 29%—about double the value of 14% obtained before,
showing that in reality we should not be at all surprised that there are nine such
pairs in the cytochrome c sequence.

A point that we glossed over at the beginning is that five out of the nine pairs
found are pairs of the same kind of amino acid, lysine, i.e. there are five KK pairs.
Even if there is no reason to be surprised that there are nine pairs of all kinds,
ought we still to be surprised that there are five that are the same? We now need
to apply the logic a bit differently from before, but it is the same sort of thinking.
Without knowing the actual frequency of lysine residues in proteins, we can guess
that it will be around 5%. So, the probability that any particular pair of residues
is KK is about 0.05^2, or 0.0025, or $1 - 0.0025 = 0.9975$ that it is not KK. By
the same logic as before, the probability that *neither* residues 1 and 2 *nor* residues 2
and 3 are KK is $0.9975 \times 0.9975 = 0.995$, and the probability that KK occurs
nowhere in the whole sequence is $0.9975^{103} = 0.77273$, and the probability that
KK occurs at least once is $1 - 0.77273 = 0.22727$. From here on, the calculation
is just the same as before, with 0.77273 instead of 0.0051 at the outset, and
$\frac{0.0025}{0.9975}$ as the multiplier in each step, and the results are as shown in
Table 8.2. (The 1.00000 in the bottom line is, of course, not exact: the probability of observing more than five KK pairs is indeed small, but it is not zero.) This
result is very different from the first one, we now see that observing five KK pairs
in a sequence of 104 is a very surprising event, one that would not be expected on
a purely chance basis, and so it is certainly worthwhile asking the question of how
it might be explained in terms of the biological role of cytochrome c and the chemical properties of lysine.

But wait: what is special about lysine? We have selected it for study here
because we already noticed that KK was the most frequent pair. However, that is
cheating, unless there was any particular reason for thinking about lysine *before* we
looked at the sequence. Rather, we should be asking what is the probability that
any dipeptide should occur five or more times. Roughly speaking, the same result
applies to any other dipeptide: the probability of five or more AA pairs is also
0.00001, as is that of five or more GG, VV, etc. As there are 20 possible pairs of
identical amino acids, this makes the probability that *none* of them occurs five or
more times about 20×0.00001, or 0.002, so instead of a 1 in 10000 event we are
looking at a 1 in 500 event. This is still rare enough to be interesting, but not so
rare as to be amazing.

I emphasize (especially in case an expert statistician should happen to read this
chapter) that the calculations we have been doing in this section have not been rigorously accurate, but that has not been the purpose. The idea has not been to
make rigorously correct probability estimates—computer programs written by

Table 8.2. Probabilities of observing specified numbers of pairs of lysine residues in a sequence of 104 amino acids.

Number of pairs	Probability	Cumulative probability	Factor for the next line
0	0.772 73	0.772 73	$(0.0025 \times 103)/(0.9975 \times 1)$
1	0.199 48	0.972 21	$(0.0025 \times 102)/(0.9975 \times 2)$
2	0.025 50	0.997 71	$(0.0025 \times 101)/(0.9975 \times 3)$
3	0.002 15	0.999 86	$(0.0025 \times 100)/(0.9975 \times 4)$
4	0.000 13	0.999 99	$(0.0025 \times 99)/(0.9975 \times 5)$
5	0.000 01	1.000 00	$(0.0025 \times 98)/(0.9975 \times 6)$

The method of calculation is the same as in Table 8.1, except that instead of 0.0051 we start with $0.77273 = 0.9975^{103}$, to allow for the fact that the probability that a pair of residues taken at random is not a pair of lysine residues is 0.9975.

professionals exist for that—but to show how some clear thinking about probabilities can allow one to form a rough idea of whether an observation is sufficiently interesting to justify doing some more exact calculations. I have also tried to give an idea of how to assess the effect of making incorrect assumptions or approximations, and in particular the direction of the bias introduced by these.

8.8 Problems

8.1 Measurements of the blood-glucose concentrations of eight people yielded the following values (all in mM): 4.17, 6.04, 5.02, 5.17, 0.85, 3.96, 5.33, 4.86. Determine the following averages:

(a) Arithmetic mean,   (b) Geometric mean,
(c) Harmonic mean,   (d) Median,
(e) Mid-range,   (f) Mode

8.2 For the values given in question (8.1), how would you interpret (as a first hypothesis) the value of 0.85 mM if

(a) all values were reported by the clinical laboratory of a hospital?

(b) all values were measured by students in a class experiment?

8.3 Of the arithmetic mean and median calculated in question (8.1), which would you expect to change more if the observation of 0.85 mM were omitted from the sample? (First try to answer this *without* recalculating; then use your calculator to see if you were right.)

8.4 For the values of question (8.1), calculate the following measures of dispersion:

(a) sample variance,   (b) sample standard deviation.

8.5 (a) Show that the formula $n\Sigma(x-\bar{x})^2$ gives exactly the same result as $n\Sigma x^2 - (\Sigma x)^2$, where all summations are for all n values in a sample and $\bar{x} = \dfrac{\Sigma x}{n}$ is the arithmetic mean.

(b) Considered as aids for computing sample variances or other statistical results such as least-squares estimates of parameters, what advantage does each formula have over the other?

(c) Use the second formula to recalculate your solution to problem (8.4a) and confirm that it is the same.

8.6 For the values of question (8.1), estimate the following measures of dispersion:

    (a) population variance,   (b) population standard deviation

    (c) standard error of the mean

8.7 Suppose that a particular value needs to be known with a precision of $\pm 2\%$ if it is to be useful, but it cannot be measured with better than $\pm 7\%$ when results from 20 experiments are combined. How many experiments would need to be made to provide the required precision if there is no change in the methods used?

8.8 Analysis of the lengths of the tails of 100 adult male rats yields a mean length of 9.2 cm and a standard deviation of 1.3 cm. Assuming a normal distribution of lengths, 95% of the lengths would be expected to lie between what limits?

8.9 A stretch of non-coding DNA of length 1000 base pairs is found to have equal frequencies of the four bases A, C, G, and T. It is found to contain eight examples of stretches with five successive bases the same, counting overlapping stretches as many times as possible, e.g. a stretch of AAAAAAA would be counted as three stretches of AAAAA. Is this a surprising observation?

Notes and solutions to problems

1.1   (a) $3x+2y+2z+2xy$; (b) $2-3x-3y$; (c) $-4x-13y$; (d) $3A-3B+3C$; (e) $-x+3y$; (f) $7x+10y$; (g) $3x-y+z$; (h) $9x+2y+z$; (i) $5a-b+c+3d$; (j) $9p-4q-2r$, (k) $5x^2-4x-2xy$.

1.2   (a) $2A^2+3AC-6AB-9BC$; (b) $2x^2+9xy+4y^2$; (c) $-8x^2+8xy+16y^2$; (d) $6xy-3xz+2y^2-yz$; (e) $4a^2+5ab+ac+12ad-6b^2+2bc-9bd+3cd$; (f) $-20p^2+19pq+7pr-3q^2-8qr+3r^2$; (g) $16-x^2$; (h) $4+4x+x^2$.

1.3   (a) $2(2x+y-3z)$; (b) $(p+q)(p+q)$ or $(p+q)^2$; (c) $(p-q)(p-q)$ or $(p-q)^2$; (d) $(p-2q)(p-2q)$ or $(p-2q)^2$; (e) $(3x+2y)(3x+2y)$ or $(3x+2y)^2$; (f) $(A+B)(A-B)$; (g) $(2u+3v)(2u-3v)$.

1.4   (a) $(x+3y)/[2(y-2x)]$; (b) $a-2b$; (c) no simplification possible; (d) $(u-v)/(u+v)$; (e) $5+2x$.

1.5   (a) $25/100$; (b) $60/100$; (c) $350/100$; (d) $30/100$; (e) $16/100$.

1.6   (a) $21/25$; (b) $1/2$; (c) $9/4$; (d) $3/17$ (no simplification possible); (e) $3/7$.

1.7   (a) $(15+14)/(7\times5)=29/35$; (b) $(15-8)/(8\times5)=7/40$; (c) 1; (d) $13/77$; (e) $81/90=9/10$; (f) $(1\times5\times7+4\times3\times7-3\times3\times5)/(3\times5\times7)=(35+84-45)/105=74/105$.

1.8   (a) $[x^2+(y+1)(y+5)]/[x(y+5)]$; (b) $[(a+3)(b-2)+(a+5)(b+1)]/[(a+5)(b-2)]$; (c) $[Va+ka(K+a)]/(K+a)$.

1.9   (a) $835/100=167/20$; (b) $-321/10$; (c) $3/4$; (d) $11/8$.

1.10   (a) 0.571; (b) 0.615; (c) 3.143; (d) 0.650.

1.11   (c) 3.142 (the others would be the same as before).

1.12   (a) $8+0.8=8.8$; (b) $8-0.8=7.2$; (c) $15+15=30$; (d) $24+4.8=28.8$, followed by $28.8-5.76=23.04$.

1.13 (a) 1.741×10^1; (b) 3.17×10^{-2}; (c) 1.5218×10^4; (d) 1.47×10^{-4};
(e) 3.1585×10^{-3};   (f) $3 \times 10^3 \times 2 \times 10^{-1} = 6 \times 10^{3-1} = 6 \times 10^2$;
(g) $3 \times 10^3 + 2 \times 10^2 = 3 \times 10^3 + 0.2 \times 10^3 = (3+0.2) \times 10^3 = 3.2 \times 10^3$.

1.14 (a) 17.41×10^0 (which would normally be written just as 17.41);
(b) 31.7×10^{-3}; (c) 15.218×10^3; (d) 0.147×10^{-3}; (e) 3.1585×10^{-3}.

1.15 (a) 67 320; (b) 0.021 55; (c) 0.018 34; (d) 418 500.

1.16 If $(x+3)=0$, i.e. if $x = -3$.

1.17 (a) 26 is divisible by 2 (it ends in 0, 2, 4, 6 or 8); (b) 4345 is divisible by 5
(it ends in 0 or 5) and by 11 $(4-3+4-5=0)$; (c) 105 is divisible by 3
$(1+0+5=6$, which is divisible by 3) and by 5 (it ends in 0 or 5); (d) 1179
is divisible by 3 and by 9 $(1+1+7+9=18$; $1+8=9$, which is divisible
by 3 and by 9); (e) 4649 is not divisible by any of the numbers proposed
(it is actually a prime number, but that is not obvious).

1.18 For the first card, there are 4 ways it could be an ace and 48 ways it could
be something else, so there is a probability of 48/52 that it is not an ace;
if the first card is not an ace, there are 4 ways the second could be an ace
and 47 ways it could be something else, so there is a probability of 47/51
that it is not an ace. Continuing in this way, the probability that none of
the first 13 cards in the hand (i.e. the whole hand) is an ace is
$(48 \times 47 \times 46 \times \cdots \times 36)/(52 \times 51 \times 50 \times \cdots \times 40) = (39 \times 38 \times 37 \times 36)/$
$(52 \times 51 \times 50 \times 49) = 0.304$.

1.19 Once the first card is dealt, there are 12 ways the second could be of the
same suit and 39 ways it could be different; hence the probability the first
two cards are in the same suit is 12/51. By a similar argument to that in
solution (1.17), the probability that the third is in the same suit is 11/50,
and so on, giving the probability that all 13 are in the same suit as
$(12 \times 11 \times 10 \times \cdots \times 1)/(51 \times 50 \times 49 \times \cdots \times 40) = 6.30 \times 10^{-14}$.

1.20 Taking reciprocals of the four values obtained, we have: (a) 2.68×10^{-6} M,
or $2.68\,\mu\text{mol}\,\text{L}^{-1}$; (b) $0.118\,\text{M}$ or $11\,800\,\mu\text{mol}\,\text{L}^{-1}$; (c) $360\,\text{M}$ or
$3.6 \times 10^8\,\mu\text{mol}\,\text{L}^{-1}$; (d) $1.78 \times 10^{-4}\,\text{M}^{-1}$. Of these, only the first is in a
reasonable range to produce a measurable effect at concentrations around
those given; (b) and (c) give values far too large; (d) has units that cannot be
correct.

1.21 150 residues correspond to a molecular mass of about $15\,000\,\text{Da}$ or
$15\,\text{kDa}$. The observed value is about four times greater than this, i.e. it is
about the value expected for a tetramer.

2.1 (a) $1+3+2+7+6+9=28$; (b) $3+2=5$; (c) $1 \times 1 + 2 \times 3 + 3 \times 2 + 4 \times$
$7 = 1+6+6+28 = 41$; (d) $1 \times 1 + 3 \times 3 + 2 \times 2 + 7 \times 7 + 6 \times 6 + 9 \times 9 =$
$1+9+4+49+36+81 = 180$; (e) $1/1 + 3/2 + 2/3 + 7/4 + 6/5 + 9/6 =$
$1+1.5+0.667+1.75+1.2+1.5 = 7.617$; (f) $1+3+2+7+6+9=28$

(if the limits are not written explicitly, the summation includes all available values).

2.2 The dimensions of [ox] and [red] cancel from the fraction, which is thus a dimensionless number, and its logarithm is likewise dimensionless. The last term, therefore, has units of RT/nF, i.e. $J \, mol^{-1} \, K^{-1} \, K \, coulomb^{-1} mol$, or $J \, coulomb^{-1}$, or $J \, A^{-1} \, s^{-1}$. These are the same as volts, the units of the two potentials. The equation is thus consistent.

2.3 (a) Inconsistent: K_m and [A] in the denominator are concentrations, but $[I]/K_i$ is a pure number; (b) consistent; (c) inconsistent: the slope should have the dimensions of a concentration, i.e. a rate divided by a rate and multiplied by a concentration, but $-1/K_m$ is a reciprocal concentration.

2.4 (a) $8+2+4=14$; (b) $8+6/7=8+0.857=8.857$; (c) $14/7=2$; (d) $8+9=17$; (e) $8\times9=72$; (f) $2^{3^2}=2^9=512$; (g) $8^2=64$.

2.5 $M_n=72\times(27\times1+17\times2+1.3\times4)/(27+17+1.3)=72\times(27+34+5.2)/45.3=72\times66.2/45.3=105\,kDa$; $M_w=72\times(27\times1^2+17\times2^2+1.3\times4^2)/(27\times1+17\times2+1.3\times4)=72\times115.8/66.2=126\,kDa$; $M_z=72\times(27\times1^3+17\times2^3+1.3\times4^3)/(27\times1^2+17\times2^2+1.3\times4^2)=72\times246.2/115.8=153\,kDa$. Although the definition refers to *numbers* of mol, we can use concentrations because they are proportional to one another and the constant of proportionality cancels from the fractions. Notice how the small proportion of tetramer has very little effect on the calculation of M_n but a large effect on the calculation of M_z.

2.6 (a) The abscissa axis is not labelled; (b) units are not given for either axis; (c) the upper line fits the data very badly (all of the points are on the same side of it); (d) the figure is dominated by a meaningless uninformative caption; (e) the lines are drawn with different thickness; (f) the points for the two experiments are illustrated with symbols of the same sort; (g) the symbols vary in size; (h) one line continues outside the plotting range; (i) the meaningless shadowing of the whole plot interferes with reading the abscissa scale; (j) the grid overlaying the data is too visible; (k) different sizes of type are used for the numbers on the two axes; (l) 75% of the plotting area contains no data; (m) there is an unexplained change of scale on the ordinate axis.

2.7 (a) 23.9; (b) 0.4925; (c) $1.3359\times0.0467=0.0624$; (d) 9.2.

2.8 (a) 8 (**23.007 050**: here and in the other answers the significant digits are shown in **bold type**); (b) 4 (0.007 **050**); (c) 3 (**135** 000); (d) 4 (**1.350** $\times 10^5$); (e) 4 (**10.37**).

3.1 (a) 27; (b) $1/16=0.0625$; (c) 3; (d) $1/\sqrt{4}=1/2=0.5$; (e) 1; (f) 0.001; (g) $3.29^{6-4-3+1}=3.29^0=1$; (h) $-2+0.37=-1.63$; (i) $4\times3\times2\times1=120$; (j) $3^3=27$; (k) 2 (in biochemical usage the symbol log means $\log_{10}$); (l) 3 (because $2^3=8$); (m) $10^3=1000$.

3.2 $e^{0.16} \approx 1 + 0.16 = 1.16$ (first two terms of the infinite series); $e^{-0.077} \approx 1 - 0.077 = 0.923$. The accuracy of the approximation can be judged from the fact that the values correct to 4 places of decimals are 1.1735 and 0.9259, respectively.

3.3 (a) $e^x \approx 1 + x$; hence $\ln(e^x) \approx \ln(1 + x)$, i.e. $x = \ln(1 + x)$; (b) 0.116 (0.1098); (c) -0.017 (-0.0171); (d) 0.041 (0.0402); (e) -0.112 (-0.1188). In each case, the values given by the approximate formula are given first, followed in parenthesis by values correct to 4 places of decimals.

3.4 (a) $\ln 4 = \ln 2^2 = 2 \times \ln 2 = 1.386$; (b) $\ln 0.2 = -\ln 5 = -1.609$; (c) $\ln 27 = \ln 3^3 = 3 \times \ln 3 = 3.297$; (d) $\ln 0.6 = \ln(0.2 \times 3) = \ln 0.2 + \ln 3 = -1.609 + 1.099 = -0.510$; (e) 0; (f) $\exp(-0.693) = 1/\exp(0.693) = 1/2 = 0.5$; (g) $\exp(1.099) = 3$; (h) $\exp(2.708) = \exp(1.099 + 1.609) = \exp(1.099) \times \exp(1.609) = 3 \times 5 = 15$; (i) $\exp(0.337) = \exp(1.946 - 1.609) = \exp(1.946)/\exp(1.609) = 7/5 = 1.4$.

3.5 Ratio $= \exp(-E_1/RT)/\exp(-E_2/RT)$, where E_1 and E_2 are the two energies. This ratio $= \exp[(E_2 - E_1)/RT] = \exp[0.04/(8.31 \times 298)] = \exp(0.000016) = 1.000016$. [The approximation used in (3.2) is virtually exact for such a small number.]

3.6 The mid-point potential is $+0.100$ volt at pH 0; it decreases linearly by 0.060 volt per pH unit till about pH 3.3, then curves gently to a smaller slope and decreases by 0.030 volt per pH unit for pH values above about 5.3.

4.1 In all cases the answers given are values of dy/dx: (a) $3x^2$; (b) $0.5/x^{1/2}$; (c) $5e^x$; (d) $5e^{5x}$; (e) 0 (e^3 is a constant); (f) $3/x$; (g) $0.5(x^{-1/2} - x^{-3/2})$; (h) $(1/x) - 4x$; (i) $x^2 e^x + 2xe^x = x(x + 2)e^x$ (derivative of a product); (j) $3x^2/x^3 = 3/x$ (function of a function); (k) $0.5 \times 2x(x^2 + 3)^{-1/2} = x/(x^2 + 3)^{1/2}$; (l) $[(x - 1) - (x + 1)]/(x - 1)^2 = -2/(x - 1)^2$ (derivative of a ratio).

4.2 (f) and (j) because $3 \ln x = \ln(x^3)$.

4.3 (a) $dy/dx = 2x - 8 + 6/x = 0$ if $x = 1$ $(y = 0)$ or if $x = 3$ $(y = -1.41)$. $d^2x/dy^2 = 2 - 6/x^2$, which is -4 (negative) when $x = 1$ and 1.33 (positive) when $x = 3$. Hence $(1, 0)$ is a maximum and $(3, -1.41)$ is a minimum.

(b) By the same analysis, there is a minimum at $(4, 12)$ and a maximum at $(6, 8)$. Notice that even though there is only one minimum and only one maximum the value of y at the minimum is larger than the value at the maximum. What happens in the vicinity of $x = 5$?

4.4 $dv/da = [(K_m + a)V - Va]/(K_m + a)^2 = K_m V/(K_m + a)^2$, hence $dv/d\ln a = a\, dv/da$ (function of a function) $= K_m Va/(K_m + a)^2$. Differentiating this a second time with respect to a gives $\dfrac{d}{da}\dfrac{dv}{d\ln a} = \dfrac{(K_m - a)K_m V}{(K_m + a)^3}$, which is zero if $a = K_m$. Hence there is a point of inflection at $a = K_m$, and at this point the plot of v against $\ln a$ has slope $K_m V K_m/(K_m+K_m)^2 = V/4$ (K_m cancelling). As $\ln a = 2.303 \log a$, the corresponding slope of Michaelis and Menten's plot of v against $\log a$ is $2.303 V/4$, i.e. $0.576 V$.

4.5 $dv/da = K_m V/(K_m+a)^2$ (see previous solution). (a) V/K_m; (b) $0.25 V/K_m$; (c) approaches 0.

4.6 (a) Writing $\bar{y}$ as a, $1/K_1$ as b and K_2 as c the equation can be expressed more simply as $y = a(1+bh+ch^{-1})^{-1}$. Hence $dy/dh = -a(b-ch^{-2})/(1+bh+ch^{-1})^2$, which is zero if $b-ch^{-2}=0$, i.e. if $h=(b/c)^{0.5}$, or, replacing b and c by their definitions, at $h=(K_1 K_2)^{0.5}$. (b) Taking logarithms, $\log h = 0.5(\log K_1 + \log K_2)$ i.e. $\text{pH} = (pK_1+pK_2)/2$.

4.7 With $K=1$ and $h=2$ the equation is $Y = x^2/(1+x^2)$, hence $Y' = 2x/(1+x^2)^2$, $Y'' = (2-6x^2)/(1+x^2)^3$. When $x=0$, $Y=0$, $Y'=0$, and $Y''=2$. When x is very large, Y approaches $x^2/x^2=1$ and Y' and Y'' both approach zero. However, Y' is positive at all positive x whereas Y'' changes from positive at low x to zero at $6x^2=2$, i.e. at $x=3^{-1/2}=0.577$, and is negative at higher values of x. Thus the curve is 'sigmoid' (S-shaped) with a point of inflection at $x=0.577$, $Y=(1/3)/[1+(1/3)]=0.25$.

4.8 $1-Y = (1+Kx^h-Kx^h)/(1+Kx^h) = 1/(1+Kx^h)$, so $Y/(1-Y) = Kx^h$. Thus, $\log[Y/(1-Y)] = \log K + h\log x$, so a plot of $\log[Y/(1-Y)]$ against $\log x$ is a straight line of slope h.

4.9 $V = (dp/dt) - K_m(-dp/dt)/(a_0-p)$. Writing dp/dt as v and a_0-p as a this is $V = v + K_m v/a$, which can be rearranged to $v = Va/(K_m+a)$, i.e. the Michaelis–Menten equation.

5.1 In all cases α and A are constants of integration: (a) $\frac{1}{3}x^3 + \frac{3}{2}x^2 + x + \alpha$;

(b) $\frac{1}{2}x^2 + \ln x + \alpha$; (c) $\frac{1}{3}\ln(2+3x) + \alpha$ or $\frac{1}{3}\ln A(2+3x)$ (begin by defining

$u = 2+3x$); (d) $-\frac{1}{3}\exp(-3t) + \alpha$; (e) define $u=2-x$, $x=2+u$, $du=-dx$,

then $\int \frac{3(2+u)du}{u} = 6\ln u + 3u + \alpha = 6\ln(2-x) + 3(2-x) + \alpha = 6\ln(2-x) - 3x + \beta$ (where $\beta = \alpha+6$ is a different constant).

5.2 (a) Evaluate $3x^2/2$ at $x=5$ and $x=0$, and subtract the latter from the former: $(3 \times 25)/2 - 0 = 37.5$; (b) $\ln 2 - \ln 1 = 0.693 - 0 = 0.693$; (c) $1^3 - (-1)^3 = 1+1 = 2$; (d) $[2^4 - (-2)^4]/4 = (16-16)/4 = 0$. In this last example, the zero area under the curve between $x=-2$ and $x=2$ is explained by the fact that areas below the x-axis count as negative. In this case, the area from $x=-2$ to $x=0$ is -4, which is cancelled exactly by the area of $+4$ from $x=0$ to $x=2$.

5.3 $y = \dfrac{5}{(3x+1)(x+2)} = \dfrac{A}{3x+1} + \dfrac{B}{x+2}$. Putting $x=0$, we have $\dfrac{5}{2} = \dfrac{A}{1} + \dfrac{B}{2}$

and with $x=1$ we have $\dfrac{5}{12} = \dfrac{A}{4} + \dfrac{B}{3}$, i.e. $5=2A+B$ and $5=3A+4B$.

Hence, $A = 3$, $B = -1$. So, $\int y \, dx = \ln(3x+1) - \ln(x+2) + \alpha$, which may alternatively be written as $\ln\left[\dfrac{A(3x+1)}{x+2}\right]$.

5.4 $[B(C+Dx) + (AD - BC)\ln(C+Dx)]/D^2 + \alpha$.

5.5 $a_0 = a + p$, hence $a = a_0 - p$, so $\dfrac{dp}{dt} = \dfrac{V(a_0 - p)}{K_m + a_0 - p}$ and separating the two variables this may be written as $\int \dfrac{(K_m + a_0 - p)dp}{a_0 - p} = \int V \, dt$ or, more simply, as $\int \dfrac{K_m \, dp}{a_0 - p} + \int dp = \int V \, dt$. The second and third integrals are trivial, and the first is of the sort considered, for example, in problem (5.1c). The solution is thus $-K_m \ln(a_0 - p) + p = Vt + \alpha$. The condition $p = 0$ when $t = 0$ gives $\alpha = -K_m \ln a_0$ and hence $Vt = p + K_m \ln[a_0/(a_0 - p)]$. Notice that this is the equation with which problem (4.9) started.

5.6 Separation of the variables gives $\int \dfrac{[K_m(1 + p/K_p) + a_0 - p]dp}{a_0 - p} = \int V dt$, which can be written as $\int \dfrac{K_m \, dp}{a_0 - p} + \int \dfrac{(K_m/K_p)p \, dp}{a_0 - p} + \int dp = \int V dt$. Of these four integrals, all but the second are identical to those in problem (5.5). Recognizing that K_m/K_p is just a constant, the second integral has the form of the integral in problem (5.1e), and the complete solution, after evaluating the constant in the same way as in problem (5.5), is $Vt = (1 - K_m/K_p)p + K_m(1 + a_0/K_p)\ln[a_0/(a_0 - p)]$.

From this equation, it follows that a plot of $t/\ln[a_0/(a_0 - p)]$ against $p/\ln[a_0/(a_0 - p)]$ gives a straight line with slope $(1 - K_m/K_p)/V$ and intercept $K_m(1 + a_0/K_p)/V$ on the ordinate.

5.7 $\int \dfrac{dI}{I} = -kc \int dx$, hence $\ln I = -kcx + \alpha$. If $I = I_0$ when $x = 0$ then $\alpha = \ln I_0$, so $\ln(I/I_0) = -kcx$, or $\ln(I_0/I) = kcx$, or $\log(I_0/I) = kcx/2.303$. Thus $A = k/2.303$.

5.8 Setting the sum of the two expressions to zero and cancelling the common factor AD gives $\dfrac{\omega^2 x(1 - \bar{V}\rho)M_r c}{RT} - \dfrac{dc}{dx} = 0$. Separating the variables and integrating gives $\ln c = \dfrac{\omega^2 x^2(1 - \bar{V}\rho)M_r}{2RT} + \text{constant}$. Thus, a plot of $\ln c$ against x^2 is a straight line whose slope can be used to calculate M_r (all of the other constants that appear in the expression for the slope being known quantities).

5.9 (a) With four strips we just consider alternate values, so $A = [0.04 + 4(2.53 + 5.75) + 2 \times 7.64 + 1.62] \times 2/3 = 33.37$. With eight strips we consider all the values in the range, so $A = [0.04 + 4(0.51 + 5.81 + 8.11 + 2.69) + 2(2.53 + 7.64 + 5.75) + 1.62] \times 1/3 = 33.99$. (b) 18.51 with four strips, 18.59 with eight. Notice that even with as few as four strips the results are almost as accurate as those with eight. With a more complicated curve, the agreement would be less satisfactory.

6.1 (a) $x=y-3$; (b) $x=\pm[(y-5)/2]^{1/2}$; (c) $x=0.2\exp(y)$; (d) $x=\frac{1}{3}\ln(0.125y)$;
(e) $x=(y+4)/(y-1)$; (f) $x=4/(3-y)$; (g) $x=-1\pm(5+y)^{0.5}$;
(h) $x=(y+2)/(y-2)$; (i) $x=[y-7\pm(25-2y+y^2)^{0.5}]/6$.

6.2 Equating the two rates, $v_1=V_2[B]/(K_{m2}+[B])$; hence $K_{m2}v_1+[B]v_1=V_2[B]$; $K_{m2}v_1=(V_2-v_1)[B]$; $[B]=K_{m2}v_1/(V_2-v_1)$. This becomes infinite if $v_1=V_2$, and negative if $v_1>V_2$; hence a steady state is impossible if $v_1\geqslant V_2$.

6.3 (a) $x=-(1\times2-1\times1)/(2\times1-1\times1)=-1/1=-1$, $y=(1\times1-4\times1)/1=-3/1=-3$.

(b) Singular, as the second equation can be obtained by multiplying the first by 2 and rearranging.

(c) $x=-6.21, y=11.47$.

(d) Singular, as the second equation can be obtained (approximately) by multiplying the first by 2.145. If the coefficients are treated as exact, the equation can be solved to give $x=2.18, y=-1.26$, but this solution is very unstable and would become $x=2.53, y=-2.57$, for example, if the coefficient of y in the second equation was 0.629 instead of 0.63. It is more realistic to treat the coefficients as accurate to two decimal places, and, in this case, the equations are singular. Equations that are nearly singular are said to be *ill-conditioned*.

(e) $\hat{a}=(\Sigma x^2\Sigma y-\Sigma x\Sigma xy)/[n\Sigma x^2-(\Sigma x)^2]$; $\hat{b}=(n\Sigma xy-\Sigma x\Sigma y)/[n\Sigma x^2-(\Sigma x)^2]$.

6.4 (a) $\Sigma x^2=1^2+2^2+3^2+4^2=1+4+9+16=30$; $\Sigma y=1.27+2.47+3.62+5.08=12.44$; $\Sigma x=1+2+3+4=10$; $\Sigma xy=1\times1.27+2\times2.47+3\times3.62+4\times5.08=37.39$. So the denominator $n\Sigma x^2-(\Sigma x)^2=4\times30-10^2=20$; $\hat{a}=(\Sigma x^2\Sigma y-\Sigma x\Sigma xy)/20=(30\times12.44-10\times37.39)/20=-0.0350$; $\hat{b}=(4\times37.39-10\times12.44)/20=1.258$.

(b) The denominator would be $1\times1-1^2=0$, i.e. the equations would be singular, illustrating the general point that one cannot estimate two constants from one observation.

6.5 (a) $1\times4-3\times2=-2$; (b) $4.71\times5.11-6.43\times1.28=15.84$; (c) $-1\times(-7)-3\times(-2)=13$; (d) 0; (e) 0 (top row consists of zeroes); (f) $2.31\times0.47-(-2.22)\times1.18=3.705$; (g) 0 (top and bottom rows are identical).

6.6 (a) $3^2-4\times1\times5=-11$: negative, so no real roots.
(b) $4^2-4\times1\times7=-12$: negative, so no real roots.
(c) $3^2-4\times(-4)\times2=51$: two real roots.
(d) $(-2)^2-4\times1\times(-12)=52$: two real roots.

(e) $3x^2 + 11x - 6 = 0$, hence $11^2 - 4 \times 3 \times (-6) = 193$: two real roots.

(f) Multiplying out, we have $7x^2 + 16x + 7 = 0$, hence $16^2 - 4 \times 7 \times 7 = 60$: two real roots.

Notice that in no case is the discriminant a perfect square, indicating rational roots. This is normal is scientific problems, where rational roots occur so infrequently that it is not worthwhile wasting time trying to factorize equations.

6.7 If the final concentration of ADP is x, then the others are (by stoichiometry) [glucose 6-phosphate] $= x$, [glucose] $= 5 - x$, [ATP] $= 4.5 - x$ (all in mM). So, $230 = x^2/(5-x)(4.5-x)$, i.e. $229x^2 - 2185x + 5175$; hence, $x = [2185 \pm (2185^2 - 4 \times 229 \times 5175)^{0.5}]/(2 \times 229) = (2185 \pm 184)/458 = 5.17$ or 4.37 mM. Although both of these are positive, only one is physically possible, because $x = 5.17$ mM would give negative concentrations for glucose and ATP. Thus the final concentration of ADP is 4.37 mM.

6.8 At 460 nm, $0.63 = 1.03 \times 10^3 \times 1 \times [B] + 4.57 \times 10^3 \times 1 \times [C]$, whereas at 500 nm, $0.52 = 7.12 \times 10^3 \times 1 \times [B] + 1.43 \times 10^3 \times 1 \times [C]$. Expressing the concentrations in mM, the factors of 10^3 disappear, so the simultaneous equations become $0.63 = 1.03[B] + 4.57[C]$, $0.52 = 7.12[B] + 1.43[C]$, with solutions $[B] = 0.0475$ mM, $[C] = 0.127$ mM.

6.9 $x = 0.211$, -0.908, or -6.970.

6.10 6.56.

6.11 (a) From the usual formula, the two solutions are $x_1 = [-b + (b^2 - 4ac)^{0.5}]/2a$ and $x_2 = [-b - (b^2 - 4ac)^{0.5}]/2a$. So $x_1 + x_2 = -b/a$ and $x_1 x_2 = [b^2 - (b^2 - 4ac)]/4a^2 = c/a$.

(b) Sum $= 7$, product $= 5$.

7.1 (a) $0.5(x^2 + y^2)^{-1/2} \times 2x = x/(x^2 + y^2)^{1/2} = x/z$; (b) $z^2 = x^2 + y^2$, so $x^2 = z^2 - y^2$, and $x = (z^2 - y^2)^{1/2}$, $\left(\dfrac{\partial x}{\partial y}\right)_z = -y/(z^2 - y^2)^{1/2} = -y/z$; (c) similarly, $z/(z^2 - x^2)^{1/2} = z/y$.

7.2 (b) Both derivatives have the value $-xy/(x^2 + y^2)^{3/2} = -xy/z^3$.

7.3 (a) $\dfrac{\partial V}{\partial T} = \dfrac{R}{p}$, $\dfrac{\partial T}{\partial p} = \dfrac{V}{R}$, $\dfrac{\partial p}{\partial V} = -\dfrac{R}{p^2}$, and the product of the three is -1.

(b) $\dfrac{\partial}{\partial p}\dfrac{\partial V}{\partial T} = \dfrac{\partial}{\partial T}\dfrac{\partial V}{\partial p} = -\dfrac{R}{p^2} = -\dfrac{V}{(pT)}$. These relationships, like the similar ones in problem (7.2), are not specific to these examples but apply in general.

7.4 $S = \sum(y_i - a + bx_i)^2$; so, $\partial S/\partial a = -2\sum(y_i - a + bx_i) = -2\sum y_i + 2an + 2b\sum x_i$ and $\partial S/\partial b = -2\sum(x_i y_i - ax_i + bx_i^2) = -2\sum x_i y_i + 2a\sum x_i + 2b\sum x_i^2$ [note that $\sum 1 = 1 + 1 + \cdots + 1$ (for n terms) $= n$]. Setting both of these equal to zero for $a = \hat{a}$ and $b = \hat{b}$ produces the pair of equations that were to be solved in problem (6.3e) and they have the solution listed above.

7.5 Refer to the solution given for problem (6.3e): ignoring the summation signs, each term in the numerator of the expression for $\hat{a}$ has dimensions of $x^2 y$ and each term in the denominator has dimensions of x^2, so both subtractions are valid and their ratio has dimensions of y, as required for an intercept on the y-axis. Similar analysis applied to the expression for $\hat{b}$ shows that it also contains valid subtractions and leads to a value with dimensions of y/x, as required for the slope of a plot of y against x.

8.1 (a) Arithmetic mean $= (4.17 + 6.04 + 5.02 + 5.17 + 0.85 + 3.96 + 5.33 + 4.86)/8 = 35.4/8 = 4.425$; (b) geometric mean $= (4.17 \times 6.04 \times 5.02 \times 5.17 \times 0.85 \times 3.96 \times 5.33 \times 4.86)^{1/8} = 3.931$; (c) harmonic mean $= 8/ (1/4.17 + 1/6.04 + 1/5.02 + 1/5.17 + 1/0.85 + 1/3.96 + 1/5.33 + 1/4.86) = 3.053$; (d) the ranked values are $(0.85, 3.96, 4.17, 4.86, 5.02, 5.17, 5.33, 6.04)$, so median $= (4.86 + 5.02)/2 = 4.94$; (e) mid-range $= (0.85 + 6.04)/2 = 3.445$; (f) there is not enough information to estimate the mode, but to the extent that the data are unimodal and the arithmetic mean and median are reliable, it should be about $3 \times 4.94 - 2 \times 4.425 = 5.97$.

8.2 (a) It is likely to represent an abnormal value that should be investigated further; (b) it probably represents a measurement or calculation error.

8.3 The arithmetic mean is much more sensitive to extreme values than the median, so it should change more. This expectation is borne out by the actual changes, which are $4.425 \rightarrow 4.936$ for the mean, and $4.94 \rightarrow 5.02$ for the median.

8.4 (a) $s^2 = [(4.17 - 4.425)^2 + (6.04 - 4.425)^2 + (5.02 - 4.425)^2 + (5.17 - 4.425)^2 + (0.85 - 4.425)^2 + (3.96 - 4.425)^2 + (5.33 - 4.425)^2 + (4.86 - 4.425)^2]/8 = 2.198$; (b) $s = 2.198^{0.5} = 1.48$.

8.5 (a) $n\Sigma(x - \bar{x})^2 = n\Sigma(x^2 - 2x\bar{x} + \bar{x}^2) = n\Sigma x^2 - 2n\bar{x}\Sigma x + n^2\bar{x}^2 = n\Sigma x^2 - 2(\Sigma x)^2 + (\Sigma x)^2 = n\Sigma x^2 - (\Sigma x)^2$ (remember that $\Sigma\bar{x} = n\bar{x}$, because $\bar{x}$ is a constant).
(b) The first formula is more transparent in meaning, with a clear relationship to the way in which the variance etc. are defined; the second is more convenient for calculation, as it avoids the need to calculate all the individual differences and both sums, Σx and Σx^2, will often have been calculated already.
(c) $s^2 = (4.17^2 + 6.04^2 + 5.02^2 + 5.17^2 + 0.85^2 + 3.96^2 + 5.33^2 + 4.86^2)/8 - 35.4^2/8^2 = 21.779 - 19.581 = 2.198$.

8.6 (a) Estimated population variance $= 2.198 \times 8/7 = 2.512$; (b) estimated population standard deviation $= 2.512^{0.5} = 1.59$; (c) standard error of the mean $= 1.48/7^{0.5} = 0.60$.

8.7 Improving the standard error of the mean by a factor $7/2 = 3.5$ implies increasing the number of observations by a factor $3.5^2 = 12.25$, so $12.25 \times 20 = 245$ experiments would be needed.

8.8 About 95% of values in a normal distribution lie within 2 standard deviations of the mean, i.e. 9.2 ± 2.6 cm, i.e. between 6.6 and 11.8 cm.

8.9   The probability that base $i+1$ is the same as base i is 0.25, and the probability that bases $i+2$, $i+3$, and $i+4$ are also the same is $0.25^4 = 0.0039$, so the probability that an arbitrary stretch of five bases contains more than one kind of base is $1 - 0.0039 = 0.9961$. The probability that this is the case at all 996 possible loci is thus 0.9961^{996}, which is most easily calculated as $\exp(996 \times \ln 0.9961) = \exp[996 \times (-0.003914)] = \exp(-3.89) = 0.0203$. Continuing as in Table 8.1, the probability that there is exactly one such stretch is $0.0203 \times (0.0039/0.996) \times 996/1 = 0.0792$, etc., leading to a probability of 0.0265 that there are eight stretches, or 0.9009 that there are seven or fewer. Thus, the observation is not so unusual as to demand special investigation.

Glossary

This section provides brief definitions of standard mathematical terms, mainly ones that are used in the book, but also some that are likely to be encountered elsewhere. Terms that appear elsewhere in the Glossary are shown in italics: for example, the definition of **abscissa** contains the words *variable* and *graph* in italics, indicating that there are also entries for these terms in the Glossary. Numbers prefixed § are the numbers of sections in the book where more information may be found.

abscissa The *variable* corresponding to the horizontal direction of a *graph*, which is normally the variable on the right-hand side of the *equation* representing the line *plotted*. [§4.1]

accuracy The exactness with which a number is known, not to be confused with *precision*. [§2.8]

antilogarithm The antilogarithm of a number x is the number whose *logarithm* is x. [§3.4]

axis A line on a *graph* representing all of the points at which one of the *variables* is zero. [§4.1]

arithmetic mean The sum of a set of values divided by the number of values; it corresponds to the everyday term *average*, but is different from the *geometric mean* and the *harmonic mean*. [§8.3]

average This term from everyday life is little used in mathematics, its most common meaning being expressed by *arithmetic mean*. See also *geometric mean, harmonic mean, median, mid-range, mode*. [§8.3]

base The base of a *logarithm* is the number that must be raised to the power of the logarithm in order to produce the *antilogarithm*. Not to be confused with the unrelated ways in which the same word is used as a chemical term. [§3.5]

brackets Any of the pairs of symbols (...), [...], and {...}, used to modify the effects of the *priority rules*. [§2.2]

Briggsian logarithm *Logarithm* to the *base* 10, synonymous with *common logarithm*. [§3.5]

calculus The branch of mathematics dealing with the ratios of *infinitisemal* changes (*differential calculus*: Chapters 4 and 7) and the cumulative effects of such changes (*integral calculus*: Chapter 5).

characteristic The *integer* part of a *common logarithm*, i.e. the part before the decimal point, the opposite of *mantissa*. [§1.5]

combination A set of choices in which the order of making choices is not important, to be contrasted with *permutation*. [§1.9]

common logarithm *Logarithm* to the *base* 10. [§3.5]

complex number See *real number*. [§6.5]

composite An *integer* is *composite* if it is not a *prime number*, i.e. it has integer *factors* apart from itself and 1. [§1.7]

constant A number that does not change over the range of conditions being considered. Mathematically, it is a number that does not change at all, but in scientific uses one cannot always be so rigid. [§2.5]

constant of integration The *constant* that must be added to the expression that results from *integrating* another expression to take account of the fact that there are an infinite number of ways of integrating a *function*. [§5.2]

coordinate geometry The branch of mathematics concerned with the expression of algebraic relations as *graphs*. [§4.1]

coordinates The pair of numbers that define the location of a point on a *graph*. [§4.1]

cross-multiplication When each side of an equation consists of a *fraction*, cross-multiplication consists of multiplying the *numerator* of the left-hand side by the *denominator* of the right-hand side and equating the *product* to the product of multiplying the numerator of the right-hand side by the denominator of the left-hand side.

decimal number A conventional way of writing a number that is equivalent to writing it as the sum of a series of powers of 10. The *fraction* $\frac{3}{8}$ corresponds to the decimal number 0.375. [§1.4]

decimal places The number of digits after a decimal point. [§2.8]

definite integral An expression representing the *difference* between two *indefinite integrals* with the same *constant of integration*. [§5.2]

denominator The bottom half of a *fraction*: in $\frac{13}{27}$ the denominator is 27, and the *numerator* is 13. [§1.3]

derivative A derivative is the result of differentiating a function. It is not the same as a *differential*, and referring to a derivative as a differential is an error. [§4.2]

determinant A square array of numbers representing a property of a set of *simultaneous equations*. [§6.4]

difference The result of subtracting one number from another: the difference between 10 and 6 is $10 - 6 = 4$.

differential calculus The analysis of ratios of *infinitesimal* changes. [Chapters 4 and 7]

differential As an adjective, this refers to the differential calculus, which is concerned with the limiting values of rates of change when the changes concerned are made indefinitely small. As a noun it refers to the *limit* of an actual change, and is not the same as a *derivative*. [Chapter 4; §7.2]

differential equation An *equation* that expresses a relationship between the *derivatives* of some *variables*. [§5.6]

dispersion The tendency of experimental measurements to be different from one another (even if they are measurements of the same thing). [§8.4]

dot notation Newton's notation for representing *derivatives* with respect to time. [§4.8]

e The number (approximately 2.718) used as the *base* of natural logarithms. [§3.7]

equation An algebraic statement that two expressions (often written in terms of one or more unknown quantities) are equal in value.

exact differential A *differential* whose value is independent of the pathway taken between two states. [§7.2]

exponent In an expression such as x^5 the 5 is an exponent that expresses how many times x has to be multiplied by itself. [§3.1]

exponential An *antilogarithm* to the base e, the result of raising e to the *power* of the number whose exponential is taken. [§3.7]

factor A number a is a factor of another number b if b/a is an *integer*. For example, 3 is a factor of 24 because $\frac{24}{3} = 8$ is an integer; 7 is not a factor of 24 because $\frac{24}{7}$ is not an integer. [§1.7]

factorial The *product* of an *integer* with all of the integers less than it: the factorial of 5 (usually written as 5!) is $5 \times 4 \times 3 \times 2 \times 1 = 120$. [§3.7]

first derivative The result of *differentiating* a *function* once, it represents the *slope* of the line on a *graph* representing the function. [§4.7]

fraction An expression that shows one number or expression, the *numerator*, divided by another, the *denominator*, such as $\frac{2}{3}$, $\frac{1.8}{2.9}$, $x/(5+y)$, etc. [§1.3]

function A relationship that can be calculated from one or more numbers. If y is a function of x then a rule exists (but is not necessarily known or specified) that allows y to be calculated from x. [§2.4]

function of a function A *function* of some other function, e.g. if z is a function of y and y is a function of x then z is a function of a function. [§4.5]

geometric mean A kind of *average*, the nth root of the product of a set of numbers. [§8.3]

graph A diagram illustrating a mathematical relationship. [§2.7]

graphical solution A method of solving *equations* by examining *graphs*. [§6.6]

harmonic mean A kind of *average*, the *reciprocal* of the *arithmetic mean* of the reciprocals of a set of numbers. [§8.3]

identity A mathematical statement whose truth does not depend on the particular values of the variables in which it is expressed. For example, $(x-y)(x+y) \equiv x^2 - y^2$ is an identity because it is true regardless of the values of x and y. Although an identity may often be written with an equality sign $=$ the identity sign $\equiv$ emphasizes that it is not just an *equation*. [§5.4]

iff A spelling sometimes used in mathematical writing to convey the idea of 'if and only if'. [§1.7]

imaginary number The *square root* of a negative number. [§6.5]

increment A change in a *variable*, not necessarily positive but measured in the positive direction: i.e. if we allow an 'increase' to be negative then an increment is the same as an increase. [§4.2]

indefinite integral An *integral* written with the *constant of integration* shown as an unknown *constant*. [§5.2]

indefinite result The result of an insufficiently defined calculation, such as the division of zero by zero. [§1.6, 4.2]

inexact differential A *differential* whose value is dependent on the pathway taken between two states. [§7.2]

infinitesimal The limit of a small value as it approaches zero. [§4.2]

infinity The result of dividing a number by zero: illegal in elementary mathematics. [§1.6]

inflection, point of A point on a *graph* at which the *second derivative* of a *function* is zero. [§4.12]

integer A whole number, i.e. a number like $0, 1, 2, 3, \ldots$, etc., whether positive or negative, i.e. 10 and -10 are both integers (though in some contexts the term may only refer to positive numbers). In this book the word *integer* is used both as a noun and an adjective, to avoid any confusion with the term *integral* as used in calculus. However, in other sources you may find *integral* used in both senses and you need to deduce from the context which is meant.

integral The result of integrating an expression (see Chapter 5). In this book *integral* is not used as the adjective from *integer*, but it may be so used in other sources. [§5.1]

integral calculus The analysis of the cumulative effects of *infinitesimal* changes. [§Chapter 5].

integrating factor A factor that allows an *inexact differential* to be made into an *exact differential*. [§7.2]

intercept The point where a line on a *graph* cuts one of the *axes*. Strictly the axis ought to be specified, but if it is not the *ordinate* axis is normally *understood*.

irrational number A number that cannot be written as the ratio of two *integers*, such as the *square root* of 2. When such a number is written as a *decimal* there is no infinitely repeating sequence, e.g. $2^{0.5} = 1.414\,213\,56 \ldots$. [§3.7]

least-squares A method of fitting an *equation* to experimental observations that makes the *sum* of *squares* of *differences* between observed and calculated values as small as possible. [§7.3]

Leibniz's notation The usual way of writing symbols for *derivatives*. [§4.8]

limit 1. The value approached by a *function* when one of the *variables* approaches infinity or zero. 2. One of the two values at which a *definite integral* is evaluated. [§5.2]

linear equation An *equation* consisting of a *constant* and a term proportional to the unknown. For example, $3x - 5 = 0$ is a linear equation. [§6.1]

logarithm The logarithm of a number x is the *exponent* to which some other number, known as the *base* of the logarithm, needs to be raised in order to give x. For example, as $10^{0.301} = 2$ (approximately), 0.301 is the logarithm of 2 to the base 10. [§3.4]

mantissa The *fractional* part of a *common logarithm*, i.e. the part after the decimal point, the opposite of *characteristic*. [§3.5]

maximum In mathematical use, this term does not have its everyday meaning of 'the biggest possible'. It refers to a value of function $f(x)$ such that any change in x, however small, will produce a smaller value of the function. A mathematical *minimum* is defined similarly. [§4.9]

mean One of a class of averages. If unqualified it usually refers to the *arithmetic mean*, but see also *geometric mean* and *harmonic mean*. [§8.3]

median A kind of *average*, the middle value from a set of numbers when they are arranged in order, or the *arithmetic mean* of the middle two values if there are an even number of them. [§8.3]

mid-range A kind of *average*, the *arithmetic mean* of the largest and smallest out of a set of numbers. [§8.3]

minimum see *maximum*. [§4.10]

mode A kind of *average* that is more useful in theoretical discussions than in practical use, it represents the most probable value. In samples that are large enough for every possible value to occur several times the mode can be taken as the value that occurs most often. [§8.3]

Napierian logarithm *Logarithm* to the *base* e, synonymous with *natural logarithm*. [§3.7]

natural logarithm *Logarithm* to the *base* e. [§3.7]

negligible A term in a sum is negligible if it is small enough for the value of the sum to be virtually unchanged if the term is omitted from the sum. Note that a comparison is always implied: 0.003 may be negligible compared with 10 (10.003 is almost the same as 10), but it is certainly not negligible compared with 0.005 (0.008 is by no means the same as 0.005).

Newton's method A method of solving *equations* by means of successive approximations. [§6.7]

Newton's notation Synonymous with *dot notation*, a way of representing *derivatives* with respect to time. [§4.8]

numerator The top half of a *fraction*: in $\frac{13}{27}$ the numerator is 13, and the *denominator* is 27. [§1.3]

operator A mathematical symbol that defines a manipulation to be applied to one or more numbers or expressions. In the expression $2 + 3$ the symbol $+$ is an operator indicating that 2 and 3 are to be added together. [§2.1]

ordinate The *variable* corresponding to the vertical direction of a *graph*, which is normally the variable that appears by itself on the left-hand side of the *equation* representing the line *plotted*. [§4.1]

origin The point on a *graph* where both of the *variables* are zero; the point at which the *axes* intersect. [§4.1]

parabola A curve representing a *quadratic function*. [§4.2]

parameter A quantity that is considered as *constant* for some purposes but as *variable* for others. [§2.5]

partial derivative If a *function* of several *variables* is *differentiated* with respect to one of those variables, the others being treated as *constants*, the resulting function is a partial derivative. [§7.1]

percentage A way of expressing a *fraction* that has 100 as *denominator*. [§1.4]

permutation A set of choices where the order in which the choices are made is important, to be constrasted with *combination*. [§1.9]

plot As a verb, to draw a *graph*; as a noun, synonymous with *graph*. [§2.7]

point of inflection A point on a *graph* at which the *second derivative* of a *function* is zero. [§4.12]

precision The exactness with which a number is expressed, not to be confused with *accuracy*. [§2.8]

priority rules Rules that determine the order in which *operators* are to be applied. [§2.2]

prime factors The complete set of *prime numbers* that produce a particular number when multiplied together: 66 can be expressed in terms of its prime factors as $2 \times 3 \times 11$. [§1.7]

prime number An *integer* that has no *factors* apart from itself and 1. For example, 7 is a prime number because there are no positive integers apart from 1 and 7 that produce 7 when multiplied together; 12 is not a prime number because it can be written not only as 1×12 but also as 2×6 or as 3×4. [§1.7]

product In mathematics, the product of a set of numbers is the result of multiplying them together: the product of 2, 5 and 7 is $2 \times 5 \times 7 = 70$. As the term also has a quite different meaning in chemistry (as in the product of a reaction), care should be taken to avoid ambiguity.

quadratic equation An *equation* in which zero is equated with a *quadratic function* of an unknown quantity. [§6.5]

quadratic function A *function* consisting of a *constant*, a term proportional to a *variable* and a term proportional to the square of the same variable. [§6.5]

quotient Equivalent to *ratio*.

ratio The result of dividing one number by another: the ratio of 12 to 4 is $\frac{12}{4} = 3$.

rational number A number that can be written as the ratio of two *integers*, for example 1.8, which is not itself an integer but can be written as $\frac{9}{5}$. When written as a decimal a real number terminates in an infinitely repeating sequence, which can be zero, as in $\frac{9}{5} = 1.800\,000\,00...$, some other digit, as in $\frac{1}{3} = 0.333\,333\,33...$, or a more complex sequence, as in $\frac{2}{11} = 0.181\,818\,18...$ or $\frac{3}{7} = 0.428\,571\,428\,571\,428\,571....$ [§3.7]

real number In this book the only numbers considered are real numbers, and that is true of most applications in biochemistry. However, in more advanced work real numbers are contrasted with *imaginary numbers*, such as the square root of -1, and *complex numbers*, which are combinations of real and imaginary numbers. Unless it qualifies the word 'number' or a word that is equivalent to a number, such as 'solution' or *root*, the word 'real' has the same meaning in mathematics as in the everyday world. [§6.5]

reciprocal The result of dividing 1 by a number, e.g. 0.5 is the reciprocal of 2, because $0.5 = \frac{1}{2}$.

residual plot A *graph* showing the differences between the calculated and observed values of a measured quantity as a *function* of some other quantity. [§2.7]

root(s) The solution(s) to an equation, used especially with quadratic and higher-order equations that have more than one solution. [§6.5]

rounding Expression of a number with decreased *precision*, in such a way that the last digit retained is as *accurate* as possible: this means that it is left as in the more precise number if the first digit to be dropped is in the range 0–4 but is increased by 1 if the first digit to be dropped is in the range 5–9. For example, 3.141 59 might be rounded to 3.142, whereas 2.718 28 would be rounded to 2.718. Compare *truncation*. [§1.5]

second derivative The result of *differentiating* a *function* twice, it provides information about the curvature of the line representing the function. [§4.7]

significant figures The digits in a number that express information about its *precision*. [§2.8]

simultaneous equations A set of two or more *equations* in two or more unknowns that are simultaneously true. [§6.3]

singular A set of *simultaneous equations* is singular if the *determinant* is zero [§6.4]. The equations cannot then be solved and the problem they represent is *underdetermined*.

slope The property of a line on a *graph* that represents the *first derivative* of the *function plotted*. [§4.1]

square The result of multiplying a number or expression by itself.

square root The positive number or expression that when multiplied by itself gives some other number or expression: the square root of 16 is 4 because $4 \times 4 = 16$. It can be regarded as the root of the *equation* $x^2 = 16$. [§2.4]

standard deviation A measure of *dispersion*. [§8.4]

stationary point A general term that includes *maximum* and *minimum* as special cases. [§4.10]

sum The result of adding two or more numbers together: the sum of 1, 7, and 16 is $1 + 7 + 16 = 24$.

tangent A straight line that 'touches' a curve at a point, i.e. both lines pass through the point and have the same *slope* at that point. The same word has a different (though related) meaning in *trigonometry* that is not considered in this book. [§4.2]

trigonometry The branch of mathematics dealing with the relationships between lengths and angles (not studied in this book as it has little application in elementary biochemistry).

truncation Expression of a number with decreased precision by simply omitting the last digits. For example, 3.141 59 might be truncated to 3.141, and 2.718 28 would be truncated to 2.718. Compare *rounding*. [§1.4]

unary A unary *operator* is one that acts on a single value. [§1.2 (Box 1.2)]

understood A symbol is understood if it is implied but not written explicitly: in the expression $1 + 2 + 3 + \cdots 10$ the missing values $4 + 5 + 6 + 7 + 8 + 9 +$ are understood. In elementary work it is usually best to be explicit.

underdetermined A problem is underdetermined if not enough information is available to solve it, e.g. if the number of unknowns in a set of *simultaneous equations* is larger than the number of equations. [§6.3]

variable A number whose value changes in the range of conditions being considered. [§2.5]

variance A measure of *dispersion*. [§8.4]

weight A factor introduced into the calculation of a *mean* to allow for information about how accurately a particular component of the mean is known. [§8.3]

Index

Page numbers in *italics* refer to boxes or problems.